ELECTRONIC ENGINEERING SYSTEMS SERIES

Series Editor
J. K. FIDLER
University of York

ACTIVE RC AND SWITCHED-CAPACITOR FILTER DESIGN
T. Deliyannis and I. Haritantis
University of Patras

THE ART OF SIMULATION USING PSPICE: ANALOG AND DIGITAL
Bashir Al-Hashimi, Staffordshire University

CIRCUIT SIMULATION METHODS AND ALGORITHMS
Jan Ogrodzki, Warsaw University of Technology

DESIGN AUTOMATION OF INTEGRATED CIRCUITS
K.G. Nichols, University of Southampton

FOUNDATIONS OF BROAD BAND LINEAR ELECTRIC CIRCUIT DESIGN
Herbert J. Carlin, Cornell University
Pier Paolo Civalleri, Turin Polytechnic

KNOWLEDGE-BASED SYSTEMS FOR ENGINEERS AND SCIENTISTS
Adrian A. Hopgood, The Open University

LEARNING ALGORITHMS: THEORY AND APPLICATIONS IN SIGNAL PROCESSING, CONTROL AND COMMUNICATIONS
Phil Mars, J. R. Chen, and Raghu Nambiar
University of Durham

OPTICAL ENGINEERING
John Watson, University of Aberdeen

OPTIMAL AND ADAPTIVE SIGNAL PROCESSING
Peter M. Clarkson, Illinois Institute of Technology

PRINCIPLES AND TECHNIQUES OF ELECTROMAGNETIC COMPATIBILITY
Christos Christopoulos, University of Nottingham

The Art of

Simulation

Using

PSpice

ANALOG *and* DIGITAL

The Art of
Simulation
Using
PSpice

ANALOG and DIGITAL

Bashir Al-Hashimi

CRC Press

Boca Raton   Ann Arbor   London   Tokyo

Library of Congress Cataloging-in-Publication Data

Al-Hashimi, Bashir.
    The art of simulation using PSpice : analog and digital / Bashir Al-Hashimi
      p.   cm.
    Includes bibliographical references and index.
    ISBN 0-8493-7895-8
    1. PSpice.   2. Electric circuits—Computer simulation.   I. Title.
TK454.A454 1994
621.3815′01′1353—dc20   94-26861
                                                                  CIP

No claim to original U.S. Government works
International Standard Book Number 0-8493-7895-8
Library of Congress Card Number 94-26861
Printed in the United States of America 1 2 3 4 5 6 7 8 9 0
Printed on acid-free paper

Preface

PSpice* is a commercial circuit-simulation package that is used to analyse analogue, digital, and mixed analogue/digital circuits. The aim of this book is to teach how to perform circuit simulation using the PSpice simulator. This is achieved by a clear, step-by-step approach illustrated by worked examples. The book also deals with advanced simulation features such as sensitivity and Monte Carlo analyses, analogue behavioural modelling, and mixed analogue and digital simulation, providing comprehensive coverage of circuit simulation using PSpice.

The book is designed to appeal to two types of readers. The first is the beginner who wants to learn about the basics of circuit analysis using PSpice. This is dealt with in Chapters 1 to 6. The second is the experienced user who is already competent in using the standard PSpice, but wants to learn about the advanced simulation features. This is dealt with in Chapters 7 to 9. In either case, the book has been written so that the reader can dip in at any section for information. It is not necessary to work through the book in order, although this would be of use to real beginners. The background required for this book is a basic understanding of electronic circuits. A good example of this background can be found in References 1 and 2.

The book is not a "simplified" PSpice manual; it tackles the subject from the viewpoint of a practical engineer with real problems to solve. The PSpice analysis commands are presented in logical groupings and order to satisfy practical engineering requirements. Also, to give the reader insight into how "good" or "bad" a simulation is, a comparison between simulation and practical results is given where necessary, in particular Chapter 6.

The book has more than 60 worked examples, which are well documented and have been tested using PSpice Design Center, Version 5.3, running on a 486DX, 33MHz IBM PC. Most of the examples will run with earlier versions of PSpice, including the student and evaluation versions. The examples in Chapter 9, however, use the digital simula-

---

* PSpice is a product of MicroSim Corporation, 20 Fairbanks, Irvine, CA 92718.

tor, which requires the extended-memory version of PSpice. Note that PSpice evaluation copies can be obtained from the MicroSim Corporation free of charge.

My view of circuit simulation, as presented in this book, has been developed over the last ten years of using and developing circuit-simulation packages for research and industry. Most of the book was written when I was a senior design engineer at Matthey Electronics, and therefore, I would like to thank the management team for their help and support. In particular, I owe much to Alan Holden for his complete and thorough review of the entire book. He made significant improvements in the choice of material, the manner of presentation, and the clarification of many ideas. I am indebted to Dr. Mansour Moniri of Staffordshire University, who read and commented on draft chapters with great perception. A special thanks must go to Ken Maddock, Richard Scott, Paul Garner, and Professor Kel Fidler of York University.

I would also like to express my appreciation and gratitude to Patrick Goss of ARS Microsystems, Navin Sullivan and Susan Fox of CRC Press, Inc., and Margaret Fletcher and her team at Staffordshire University.

References

1. Bogart, T.F., *Electronics Devices and Circuits*, 2nd edition, Merrill Publishing Company, Columbus, OH, 1990.
2. Mano, M.M., *Digital Designs*, 2nd edition, Prentice Hall, Englewood Cliffs, NJ, 1991.

Contents

This book is dedicated to my wife, May, without whose help, support, and encouragement, it would not have been possible.

chapter one

Introduction

1.1 Brief history of SPICE

SPICE stands for *"Simulation Program with Integrated Circuit Emphasis"*. It is a powerful computer package used to analyse electronic circuits. *SPICE* was developed in the early 1970s in the Department of Electrical Engineering of the University of California at Berkeley, progressing through various versions, culminating in 1981 with *SPICE 2G.6*, which is the version on which most commercial circuit-simulation packages are based. The algorithms used in *SPICE* are discussed in detail in Reference 1. It should be noted that the program *SPICE* is freely available. Although *SPICE 2G.6* is a powerful and robust program, it did not gain popularity in nonacademic circles for a number of reasons. The early versions needed to run on mainframe computers and had "user-unfriendly" interfaces. It was not interactive and lacked the ability to simulate digital and mixed analogue/digital circuits. Finally, there were no component libraries, and all these features made the original *SPICE* unattractive.

A number of commercial organisations could see the potential of the core software, provided that these shortcomings were rectified, and versions are now available for use on a variety of computers ranging from PCs to mainframes. Table 1.1 shows some IBM-PC-based *SPICE* versions. Reference 2 gives a detailed list of commercially available circuit-simulation packages, including workstations and mainframe computers.

PSpice is arguably the industry standard for circuit-simulation packages with more than 17,000 copies sold worldwide to date.

1.2 Book structure

This book is based on PSpice Design Center, Version 5.3, and consists of nine chapters. The first six chapters cover all aspects of PSpice circuit simulation and analysis. Chapters 7 to 9 cover the advanced simulation features.

Table 1.1 Some Commercially Available
PC-Based SPICEs

Vendor	Simulator name
Contec Microelectronics	ContecSpice
Deutsch Research	TurboSpice
Electrical Engineering Software	Precise
Intusoft	IsSpice
Meta-Software	HSpice
MicroSim Corporation	PSpice
Spectrum Software	Micro-Cap

PSpice contains more than 20 command instructions that allow the designer to perform various circuit simulations. The book is organised in a way that is parallel to the normal development and testing of an electronic circuit. Chapter 2 deals with the description of passive components and simple active elements (diodes and transistors), and it shows how to generate the circuit in a form that PSpice understands (the "netlist"). Other key parameters, such as DC and AC sources and power supplies, are also introduced. The reader should be able to produce simulation results of simple circuits by the end of the chapter.

Chapter 3 builds on Chapter 2 and introduces the important concepts of time-domain and Fourier analysis, which are often used in measuring the performance of circuits (such as transient response). The complex active devices are dealt with in Chapter 4, where amplifiers and similar devices are described in terms of controlled sources and equivalent circuits. Having completed Chapter 4, the reader should be able to perform extensive simulations.

Complex circuits often contain repetitive circuit elements, and these may be expressed as subcircuits. Chapter 5 covers the definition and use of the subcircuit approach in circuit description. Actual op-amp models (macromodels) and their usage are discussed in Chapter 6, which also compares and contrasts the result from various simulations with the practical results. Chapters 7 to 9 discuss in detail various specific simulation topics (including sensitivity and Monte Carlo analyses, analogue behavioural modelling and mixed analogue/digital simulation).

The seven appendices that accompany this book provide the user with additional information on the advanced usage of PSpice and specific circuit-simulation problems. Appendix A lists some of the common error messages reported by PSpice when there is an error in the circuit. Appendix B shows how time-dependent input signals are generated using the program (*StmEd*). Appendix C covers the simulation of oscillator circuits, and Appendix D deals with the simulation of switched capacitor networks. The simulation of ideal and lossy transmission lines is discussed in Appendix E, and Appendix F explains the various PSpice control options. Finally, Appendix G covers the simulation of CMOS circuits.

1.3   PSpice syntax

PSpice syntax is simple because it has only two kinds of statements, description and control. Description statements tell PSpice about component types and their connections, while control statements instruct PSpice to perform particular tasks or analyses. These control statements are differentiated from the description statements by always beginning with ".".

1.4   Simulation hierarchy

The use of circuit-simulation packages is increasing because of the advantages they offer the user. At the simplest level they allow designers to verify whether their circuits function as designed. This level of simulation often represents stage one of the simulation process. Ideal component models are often assumed during this stage of simulation. The objective of the stage one simulation is to make certain that the designer has not made any fundamental mistakes in the design. The advantage of stage one is that it takes little time and effort to produce simulation results, while the disadvantage is that only ideal circuit performance is predicted. Chapters 2 through 5 deal with stage one of circuit simulations.

To improve the accuracy between the simulation and the practical results, more complex component models are needed in the simulation. Here, the user has to decide on the accuracy required in the simulation. Generally, the higher the accuracy required, the more time, effort, and better understanding of the circuit operation are required. The simulation process might also involve examining the sensitivity of the circuit to component and temperature variations. This type of simulation often represents stage two of the simulation process, which is dealt with in Chapters 6 and 7.

So far, component-level simulations have been considered. PSpice offers the designer the option to perform system-level simulation, which can be a very effective method of simulating complex circuits. This is accomplished by the use of analogue behavioural modelling and is discussed in Chapter 8. The majority of today's electronic circuits contain both analogue and digital circuitry. PSpice allows the designer to simulate the performance of such systems, and this is covered in Chapter 9.

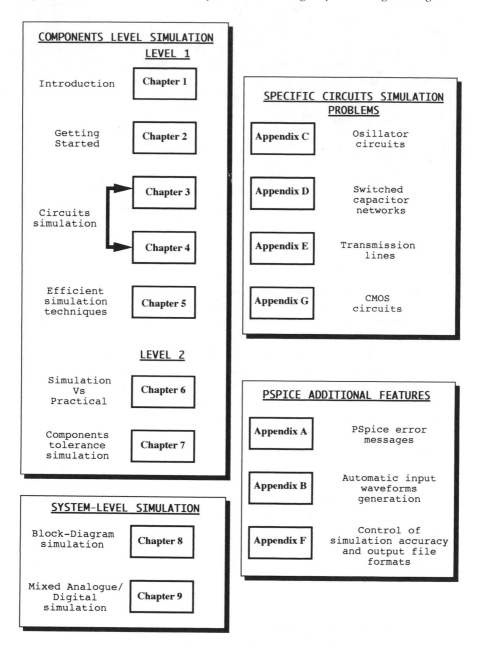

Figure 1.1 Book structure.

1.5   Chapter summary

The book has been organised as to emphasise the hierarchical view-point of component and system simulation, as illustrated in Figure 1.1. Much simulation can be carried out with a working knowledge of Chapters 2 through 4. However, later chapters build on these to enable a more complex system-level approach to be realised by the reader.

References

1. Nagel, L.W., *SPICE2: A Computer Program to Simulate Semiconductor Circuits*, ERL Memo No. ERL-M520, Electronics Research Laboratory, University of California, Berkeley, May 1975.
2. Swager, A.W., "Behavioral Models Expedite Simulation", *EDN Magazine*, November 1991. pp. 67-75.

chapter two

Circuit entry and preliminary analyses

This chapter shows how components are described for entry into PSpice. The simple voltage and current sources contained in PSpice are discussed, as well as the preliminary analyses possible. Examples of passive and active circuits are given to illustrate the various steps from component entry through to analysis and the different output options.

2.1   Introduction

In order to carry out simulation, the circuit must first be described to PSpice. Prior to circuit description, it is good practice to perform the following steps:

1. Create a schematic diagram of the circuit.
2. Give each component in the circuit a name.
3. Identify and number all the nodes in the circuit.
4. Assign each component its value or size.

2.2   Passive components description

Having performed steps 1 to 4 above, one can now describe each component in the circuit using a component description statement. The basic syntax is:

<component name> <node 1> <node 2> <value>

The parameter *<component name>* is the name of the component, and this name must start with a letter recognised by PSpice. Table 2.1 shows the letters recognised by PSpice as components. The table shows, for example, that a resistor is identified by the letter R, while a capacitor is identified by the letter C.

Table 2.1   Components Recognised by PSpice

Letter	Description
B	Gallium arsenide field effect transistor (GaAsFET)
C	Capacitor
D	Diode
J	Junction field effect transistor (JFET)
K	Coefficient of coupling for mutual inductance
L	Inductor
M	Metal oxide semiconductor field effect transistor (MOSFET)
Q	Bipolar junction transistor (BJT)
R	Resistor
S	Voltage-controlled switch
T	Transmission line
W	Current-controlled switch

Component names can be up to eight characters long, but must start with a letter recognised by PSpice. The name can contain either letters, numbers, or both. For example, the resistors in the circuit of Figure 2.1 could be named R1 and R2 or Rin and Rout, respectively. Note that each component must have a unique name. It is not essential to describe components using upper-case letters; lower-case letters would have been just as valid.

The parameters *<node 1>* and *<node 2>* show how a component is connected within a circuit. A node is defined as a point of connection between two or more circuit components. For example, the input voltage of the circuit in Figure 2.1 is at node 1, while the output voltage is at node 3. PSpice assumes the circuit ground is at node 0. It is essential for every circuit to have a ground node as other node voltages will be expressed with respect to it. PSpice accepts all real numbers as nodes,

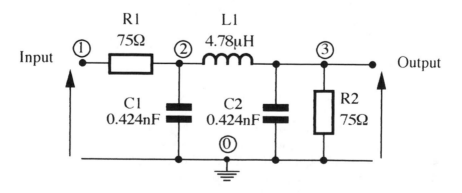

Figure 2.1   RLC circuit.

Table 2.2 PSpice Scale Suffixes

Symbol	Name	Scale	Some valid PSpice commands for component of value (4.7*scale factor)
k	kilo	10^{+3}	4.7K, 4.7k, 4K7, 4k7, 4.7E3
M	mega	10^{+6}	4.7MEG, 4.7meg, 4.7E6
G	giga	10^{+9}	4.7G, 4.7g, 4G7, 4g7, 4.7E9
T	tera	10^{+12}	4.7T, 4.7t, 4T7, 4t7, 4.7E12
m	milli	10^{-3}	4.7M, 4.7m, 4M7, 4m7, 4.7E-3
μ	micro	10^{-6}	4.7U, 4.7u, 4U7, 4u7, 4.7E-6
n	nano	10^{-9}	4.7N, 4.7n, 4N7, 4n7, 4.7E-9
p	pico	10^{-12}	4.7P, 4.7p, 4P7, 4p7, 4.7E-12
f	femto	10^{-15}	4.7F, 4.7f, 4F7, 4f7, 4.7E-15

but for clarity it is suggested that positive integers be used. For example, the resistor R1 in Figure 2.1 is connected between nodes 1 and 2. The order in which nodes are numbered is arbitrary, but each node number must be unique.

In practice, component values are often expressed in terms of scale suffix notation. For example, the values of C1 and L1 in Figure 2.1 are usually specified as 0.424nF and 4.78μH, respectively. PSpice allows the use of such notation, as shown in Table 2.2. This table shows that PSpice is very accommodating in the way component values can be expressed. Note that the symbol (μ) is described in PSpice as (u).

Using the component description statement, the circuit in Figure 2.1 is described in PSpice as follows:

```
R1  1  2  75
C1  2  0  0.424n
L1  2  3  4.78u
C2  3  0  0.424n
R2  3  0  75
```

This shows that the resistor R1 is connected between nodes 1 and 2 and its value is 75 Ω, C1 is connected between nodes 2 and 0 and its value is 0.424nF. Similarly, L1 is connected between nodes 2 and 3 and its value is 4.78μH. Note that the components have no units after their values in the listing. PSpice assumes the basic electrical units for component values, as shown in Table 2.3. If desired, the unit of a component can be included. For example, the values of C1 and L1 could have been expressed in PSpice as 0.424nF and 4.78μH, respectively, without affecting the values. This is because once PSpice recognises a valid suffix ("n" in the case of the capacitor value, and "μ" in the case of the inductor value), the remaining letters are ignored and the first letter determines the units.

Table 2.3 PSpice Default Units

Type	Unit
Resistor	ohm
Inductor	henry
Capacitor	farad

It is also possible to use the exponential (E) notation in expressing component values, where E denotes the power of 10. For example, the values of C1 and L1 in Figure 2.1 could have been specified as 0.424E–9 and 4.78E–6, respectively.

2.3   Semiconductor components description

Thus far, only passive components (resistors, capacitors, and inductors) have been described using the passive component statement. This statement allows the user to express a passive component in terms of component name, nodal connections, and component value. To describe semiconductor components, two statements are required. The first, a component description statement, defines the component type with input and output nodes. The second, a model statement, describes component characteristics in terms of a set of parameters.

As an illustration, the PSpice description of diodes and transistors will be considered here. Other semiconductor components will be covered in Chapter 6 and Appendix G.

2.3.1   Diode description

The basic form of the diode component description statement is:

$$D<name> <NA> <NK> <model\ name>$$

where D is the PSpice symbol for a diode, and <name> is the diode name, which can be up to eight characters long. Names are chosen arbitrarily by the user, for example, D1, DREF1, and DCLAMP. The parameters <NA> and <NK> are the anode and the cathode nodes of the diode, as shown in Figure 2.2.

Anode   Cathode

Figure 2.2   Diode node connections.

The parameter *<model name>* is usually the descriptive name of the diode. This name can begin with any character and can be up to eight characters long. Examples of model names are TYPE1 and LOWPOWER.

The basic form of the diode model statement is

.MODEL *<model name>* D [*model parameters*]

where the parameter *<model name>* is the name given to the diode in the component statement. The letter D is the PSpice symbol for the diode. (Note that in the .MODEL statement, the "." symbol is a necessary part of the statement line.)

The "[ ]" symbol in a PSpice statement means that the provision of parameters by the user is optional. If no parameters are supplied, PSpice will use the default values. For a diode, the values shown in Table 2.4 are used.

Table 2.4   Diode Model Parameters

Parameters	Description	Default	Unit
IS	Saturation current	1E-14	A
N	Emission coefficient	1	
ISR	Recombination current parameter	0	A
NR	Emission coefficient for ISR	2	
IKF	High-injection "knee" current	Infinite	A
BV	Reverse breakdown "knee" voltage	Infinite	V
IBV	Reverse breakdown "knee" current	1E-10	A
NBV	Reverse breakdown ideality factor	1	
IBVL	Low-level reverse breakdown "knee" current	0	A
NBVL	Low-level reverse breakdown ideality factor	1	
RS	Parasitic resistance	0	Ohm
TT	Transit time	0	s
CJO	Zero-bias p-n capacitance		F
VJ	p-n Potential	1	V
M	p-n Grading coefficient	0.5	
FC	Forward-bias depletion capacitance coefficient	0.5	
EG	Bandgap voltage (barrier height)	1.11	eV
XTI	IS Temperature exponent	3	
TIKF	IKF Temperature coefficient (linear)	0	$°C^{-1}$
TBV1	BV Temperature coefficient (linear)	0	$°C^{-1}$
TBV2	BV Temperature coefficient (quadratic)	0	$°C^{-2}$
TRS1	RS Temperature coefficient (linear)	0	$°C^{-1}$
TRS2	RS Temperature coefficient (quadratic)	0	$°C^{-2}$
KF	Flicker noise coefficient	0	
AF	Flicker noise exponent	1	

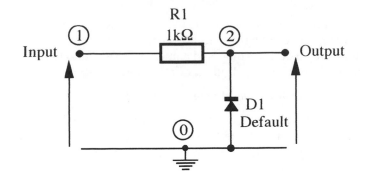

Figure 2.3   Simple diode circuit.

PSpice has a detailed and accurate model for the diode, and this can be used to model many DC and AC characteristics. The model has 25 parameters, each of which has a specific default value as shown in Table 2.4. For example, the default value of the diode saturation current (IS) is 1E-14, and the reverse breakdown voltage (BV) is infinite. The default values are chosen so that an ideal diode performance is obtained. For simple circuit analysis, it is generally sufficient to allow most model parameters to have their default values. There are situations, however, that would require accurate values for the model parameters. A detailed discussion of components model parameters will be covered in Chapter 6.

As an example, consider the circuit in Figure 2.3. The PSpice description of this circuit is

```
R1  1  2  1k
D1  0  2  Default
.MODEL  Default  D
```

The diode, D1, is connected between nodes 0 and 2, with the anode at node 0. The model name of this diode is "Default" and has been chosen arbitrarily. The diode model parameters have been set to their default values, since no model parameters were specified in the .MODEL statement of the diode.

2.3.2   *Bipolar transistor description*

The basic form of the bipolar transistor component description statement is

Q<name> <NC> <NB> <NE> <model name>

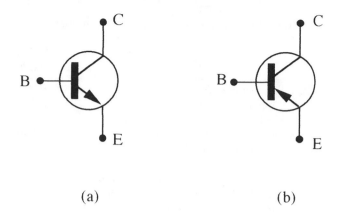

(a)   (b)

Figure 2.4   Transistor node connections: (a) NPN; (b) PNP.

where Q is the PSpice symbol for a bipolar junction transistor, and *<name>* is the transistor name, which can be up to eight characters long. Names are chosen arbitrarily by the user, and examples are Q1, Qin, and Qbias. The parameters *<NC>*, *<NB>*, and *<NE>* are the collector, base, and emitter nodes, respectively, as shown in Figure 2.4. The parameter *<model name>* is usually a descriptive name of the transistor. This name can begin with any character and can be up to eight characters long. Examples of model names are TYPE1, and TRAN.

The basic form of the bipolar transistor model statement is

.MODEL *<model name> <type>* [*model parameters*]

which is similar to the .MODEL statement of the diode. The parameter *<model name>* is the name given to the transistor specified in the component statement. The parameter *<type>* determines the transistor type, which could be either NPN or PNP. PSpice has an accurate and comprehensive model of the bipolar transistor and can be used to model many effects ranging from the current gain and capacitance junctions to noise and temperature characteristics. The transistor model has more than 40 parameters, as shown in Table 2.5. In practice, some of these transistor parameters are difficult to obtain or measure and are necessary only when a highly accurate model is essential. For simple circuit simulation, it is sufficient to allow most of the transistor parameters to have their default values. Each model parameter has a specific default value. For example, the default value of the <ideal maximum forward beta>, or (BF), is 100. If these parameters are not specified by the user using the [*model parameters*] part of the .MODEL statement, PSpice will set the parameters to their default values to give typical component performance.

Table 2.5   Bipolar Transistor Model Parameters

Parameter	Meaning	Default value	Unit
IS	p-n Saturation current	1E-16	A
BF	Ideal maximum forward beta	100	
NF	Forward current emission coefficient	1	
VAF(VA)	Forward early voltage	Infinite	V
IKF(IK)	Corner for forward beta high-current roll-off	Infinite	A
ISE(C2)	Base-emitter leakage saturation current	0	A
NE	Base-emitter leakage emission coefficient	1.5	
BR	Ideal maximum reverse beta	1	
NR	Reverse current emission coefficient	1	
VAR(VB)	Reverse early voltage	Infinite	V
IKR	Corner for reverse beta high-current roll-off	Infinite	A
ISC(C4)	Base-collector leakage saturation current	0	A
NC	Base-collector leakage emission coefficient	2.0	
RB	Zero-bias (maximum) base resistance	0	Ohm
RBM	Minimum base resistance	RB	Ohm
IRB	Current at which RB falls halfway to RBM	Infinite	A
RE	Emitter ohmic resistance	0	Ohm
RC	Collector ohmic resistance	0	Ohm
CJE	Base-emitter zero-bias p-n capacitance	0	F
VJE(PE)	Base-emitter built-in potential	0.75	V
MJE(ME)	Base-emitter p-n grading factor	0.33	
CJC	Base-collector zero-bias p-n capacitance	0	F
VJC(PC)	Base-collector built-in potential	0.75	V
MJC(MC)	Base-collector p-n grading factor	0.33	
XCJC	Fraction of Cbc connected internal to RB	1	
CJS(CCS)	Collector-substrate zero-bias p-n capacitance	0	F
VJS(PS)	Collector-substrate built-in potential	0.75	V
MJS(MS)	Collector-substrate p-n grading factor	0	
FC	Forward-bias depletion capacitor coefficient	0.5	
TF	Ideal forward transit time	0	s
XTF	Transient time bias dependence coefficient	0	
VTF	Transit time dependency on Vbc	Infinite	V
ITF	Transit time dependency on Ic	0	A
PTF	Excess phase at $1/(2\pi .TF)$Hz	0	°
TR	Ideal reverse transit time	0	s
EG	Bandgap voltage (barrier height)	1.11	eV
XTB	Forward and reverse beta temperature coefficient	0	
XTI(PT)	IS Temperature effect exponent	3	
KF	Flicker noise coefficient	0	
AF	Flicker noise exponent	1	

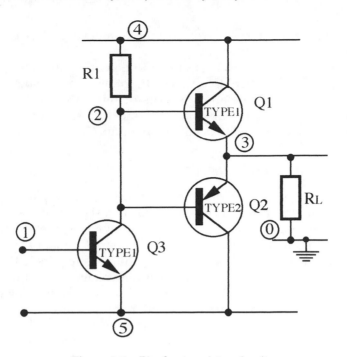

Figure 2.5 Bipolar transistor circuit.

As an example, consider the circuit in Figure 2.5. The PSpice descriptions of the bipolar transistors are

```
Q1  4  2  3  TYPE1
Q3  2  1  5  TYPE1
Q2  5  2  3  TYPE2
.MODEL  TYPE1  NPN
.MODEL  TYPE2  PNP  [BF=80]
```

The transistor Q1 has its collector, base, and emitter connected to nodes 4, 2, and 3, respectively. It is an NPN type with all the transistor model parameters set to default values, since no model parameters were specified in the first .MODEL statement. Similarly, Q3 is an NPN transistor with the same model parameters as Q1, since both transistors use the same model (TYPE1). Note that it is necessary to specify a particular type only once.

The transistor Q2 is a PNP type and has its collector, base, and emitter connected to nodes 5, 2, and 3, respectively. The transistor Q2 uses the model (TYPE2), which has all its parameters set to their default values, apart from the (ideal maximum forward beta), or (BF), parameter which is set to 80 using the [model parameters] of the .MODEL bipolar transistor statement. Note that the default value of BF is 100 (see Table 2.5).

The description of JFETs is similar to that of bipolar. PSpice identifies a JFET by the letter J (see Table 2.1), followed by its nodes (drain, gate, and source). There are two types of JFETs (N and P channel), which are described in PSpice as NJF and PJF, respectively. CMOS transistors are discussed in detail in Appendix G.

2.4   Independent sources

Thus far, the methods of describing circuits using simple active and passive components have been discussed. Test signals and power supplies can be specified in PSpice, and this allows a complete simulation of the operation of the circuit. To define an input signal to the circuit, an independent voltage and/or current source is required. The word "independent" in this sense means that the voltage across the voltage source or the current flowing through the current source is not dependent on any other circuit parameters. Independent sources are specified in PSpice in terms of a name, nodal connections, type, and value. The basic form of an independent source description statement is

<source name> <+ node> <– node> <type> <value>

where the parameter *<source name>* is the independent voltage or current source name. These sources are identified by PSpice as components starting with the letters V and I, respectively, as shown in Table 2.6. As with components, the letters V and I can be followed by any number of characters up to eight, to represent a source name. Examples of independent sources are Vin, V1, Vcc, and Isignal.

The parameters *<+ node>* and *<– node>* are the positive and negative nodes, respectively, of the independent source. The parameter *<type>* defines the type of the independent source, while *<value>* specifies its magnitude. There are three types of independent sources available in PSpice; these are DC, AC, and transient sources.

2.4.1   DC Independent source

This source is used to specify a fixed value of voltage or current. The description statement of a DC independent source is

<source name> <+ node> <– node> DC <magnitude>

where *DC* specifies a dc-type source and *<magnitude>* represents a fixed voltage or current value. This value can be positive or negative. DC

Table 2.6   Independent Sources
Recognised by PSpice

Letter	Description
V	Independent voltage source
I	Independent current source

independent sources can be used to specify power supplies and current sources, for example:

```
Vcc  6  0  DC  5V
Vee  0  7  DC  5V
Ibias  10  11  DC  3mA
```

The first statement specifies a positive power supply named "Vcc" connected between nodes 6 and 0 (ground) with a magnitude of 5V. The second statement describes a negative power supply, named "Vee", connected between nodes 0 and 7 with a magnitude of 5V. Note that it is just as valid to describe this negative power supply using the following statement:

Vee 7 0 DC –5V

The third statement describes a current source called "Ibias" connected between nodes 10 and 11 with a DC value of 3mA. If the independent source is not specified DC, PSpice will automatically assume that it is DC. This means the first statement could have been written as

Vcc 6 0 5V

without altering its meaning. PSpice also assumes default values for electrical quantities as shown in Table 2.7. This means that the value of "Vcc" in the first example could have been written as 5 without affecting the value of the voltage source. Similarly "Ibias" could have been written as 3m.

Table 2.7   PSpice Default Values
for Electrical Quantities

Type	Unit
Voltage	Volt
Current	Ampere

PSpice is capable of performing DC sweep analysis. This type of analysis allows the circuit to be tested using various DC values. This analysis is achieved using the following statement:

.DC <source name> <start value> <end value> <step value>

where the parameter *<source name>* is an independent source (voltage or current), which must already be specified using a DC independent source description statement. The parameters *<start value>* and *<end value>* define the start and the final value of the voltage or current source. The parameter *<step value>* represents the size of the step used to go from the start to the end value of the source.

A typical example of the use of DC sweep analysis is to examine the effects of varying the power supply values on circuit performance. The following statements are required:

```
Vcc  12  0
.DC  Vcc  12V  15V  1V
```

The first statement defines the power supply, Vcc, which is connected between nodes 12 and ground. The second statement instructs PSpice to sweep the value of this power supply from 12 to 15V in steps of 1V.

In some applications, there may be a requirement for multiple voltage and current sources. It is possible to simulate the effects of multiple variation by using "nested" DC sweep analysis. An illustration of this is given in Example 2.7.1.

In addition to the .DC statement, PSpice has two further statements specifically for DC operating conditions. These are the .OP (operating points) and .TF (transfer function) statements. The basic form of the .OP statement is

.OP

This statement causes detailed information about the circuit bias points to be calculated. This includes circuit node voltages, currents, and power dissipation, as well as the small signal or DC parameters of semiconductors. See Example 2.7.2 on the use of this statement. In fact, PSpice calculates by default all the circuit node voltages, whether or not there is an .OP statement.

The basic form of the .TF statement is

.TF <output variable> <input source name>

This statement causes PSpice to calculate the circuit DC transfer function, where transfer function is the ratio of the *<output variable>* to the

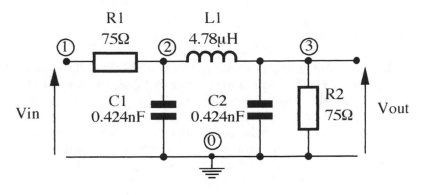

Figure 2.6 Lowpass filter circuit.

<input source name>. This statement also calculates the DC circuit input resistance at the *<input source>* and the output resistance at the *<output variable>* node. As an example, consider the circuit in Figure 2.6.

To find out the DC gain, input resistance, and output resistance of the circuit, the following .TF statement is required:

$$.TF\ V(3)\ Vin$$

where V(3) is the voltage at node 3 or the output voltage, and Vin is the input voltage. PSpice will calculate the DC gain to be 0.5 of the signal at Vin. This is because R1 = R2 = 75Ω. The input resistance at Vin will be 150Ω, while the output resistance at node 3 will be 37.5Ω. These results are included in the circuit output file (see Figure 2.7).

2.4.2   AC Independent source

This source is used when the response of a circuit over a range of frequencies is required (i.e., frequency response simulation). This is the same as using the sweep frequency signal generator. In PSpice, two statements are required to define the sweep frequency signal: an AC independent source description statement and an .AC statement. The description statement has the following form:

<source name> <+ node> <– node> AC <magnitude> <phase>

where *AC* specifies AC type source, *<magnitude>* describes the peak amplitude of the AC signal source in volts or amperes, and *<phase>* represents the phase angle of the AC source in degrees. If the magnitude value of the signal is not specified, PSpice assumes the default value of 1V. Similarly, if the phase angle is not specified, PSpice uses the default value of 0°. Examples of independent voltage and current AC sources are

```
Vin 1 0 AC 1
Isource 5 0 AC 1m
```

The first statement describes an AC voltage source called Vin connected between nodes 1 and 0, with a magnitude of 1V and a phase angle of 0°. The second statement represents an AC current source connected between nodes 5 and 0 with a value of 1mA.

Normally the phase angle specification is not useful when dealing with a single AC source. However, the phase angle might become important when a description of a complex input signal (made of a number of signals of different amplitudes and phases) is required.

The frequency range of an AC independent source is specified using the following statement:

.AC <sweep type> <N> <fstart> <fstop>

where the parameter *<sweep type>* defines how the frequency points are determined in the range of *<fstart>* and *<fstop>*. PSpice does not allow the AC analysis to start at DC; therefore, the <fstart> must be a number >0.

PSpice has two kinds of frequency sweeps, linear and logarithmic. To specify a linear frequency scale, the following statement is required:

.AC LIN <N> <fstart> <fstop>

where *LIN* specifies a linear frequency scale. The parameter *<N>* defines the number of frequency points in the linear sweep starting at *<fstart>* and finishing at *<fstop>*. As an example, the statement

.AC LIN 100 1 1K

instructs PSpice to perform an AC analysis over the frequency range of 1Hz to 1kHz with 100 frequency points in the range. There are two options available in which a logarithmic frequency scale is specified. The first is a decade scale (a tenfold increase in frequency), and the second is an octave scale (a twofold increase in frequency). To specify a decade frequency scale, the following statement is required:

.AC DEC <N> <fstart> <fstop>

where *DEC* specifies a decade logarithmic frequency scale, and *<N>* is the number of frequency points per decade. As an example, the statement:

.AC DEC 10 10k 10meg

causes PSpice to divide the frequency range into decades, and in this case there are three decades. PSpice will perform AC analyses at 31 frequency points. Similarly, to specify an octave logarithmic frequency sweep, the word *DEC* in the decade scale needs to be changed to *OCT*. As an example, the statement

.AC OCT 5 1k 16k

instructs PSpice to divide the frequency range into octaves, and in this case there are four octaves. PSpice will perform AC analyses at 21 frequency points.

2.4.3 Transient independent source

This source is used when the response of a circuit over a specified time range is required (i.e., time domain simulation). Transient sources are very powerful and flexible, allowing a large variety of complex analyses to be performed. Consequently, this topic is dealt with in detail in Chapter 3.

2.5 Output options of simulation results

So far, we have discussed the description of some passive and active circuits, as well as power supplies and input signals. PSpice always produces a text output file with an .OUT extension. The contents of this file are determined by the input file commands. This section discusses the various PSpice output options available to the user. Figure 2.7

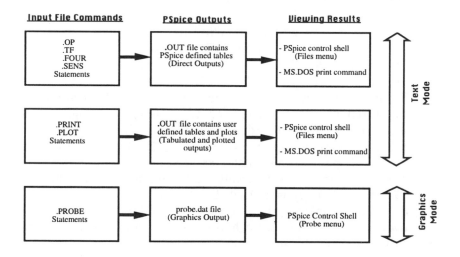

Figure 2.7 PSpice output options.

shows the various command paths possible. Two main modes are available, "text mode" and "graphic mode". Finally, the control shell (which is discussed in Section 2.6) is normally used to provide a "user-friendly" interface to PSpice for viewing both input and output files.

2.5.1 Direct outputs

Some commands produce output directly into the .OUT file. These are PSpice-defined outputs, and two of these commands (.OP and .TF) have been mentioned in Section 2.4.1. There are others as given in the direct output command box shown in Figure 2.7.

2.5.2 Tabulated and plotted outputs

If it is desired to produce the simulation results in the form of a user-defined table, then the .PRINT statement must be used. The basic form of this statement is

.PRINT <analysis type> <*output variables*>

The parameter <*analysis type*> can be DC, AC, noise, or transient analysis. The parameter <*output variables*> can be a series of variables such as node voltages and component currents. Each output variable becomes a column in the table that is stored in the circuit output file. Node voltages and component currents can be specified as magnitude, real part, imaginary part, or in decibels. Examples are

.PRINT DC V(3) I(R1)
.PRINT AC V(5,6) VR(5,6) VI(5,6)
.PRINT AC VDB(10)

The first statement prints the magnitude of the voltage at node 3, and the current through R1 of a DC analysis. Here, the voltage at node 3 is printed relative to ground (or node 0), which is the default reference node. The second statement prints the magnitude and the real (VR) and imaginary (VI) parts of the voltage between nodes 5 and 6 of an AC analysis. The third statement prints the magnitude of the voltage at node 10 of an AC analysis in decibels.

It is also possible to print out the phase and/or the group delay (derivative of phase with respect to frequency) of an AC analysis using the .PRINT statement. Examples are

.PRINT AC VP(6) VG(6)

This statement causes PSpice to print the phase and the group delay, respectively, of the voltage at node 6.

If it is desired to produce the simulation results in the form of a plot, then the .PLOT statement is required. The basic form of this statement is

.PLOT <analysis type> <output variables>

This statement is very similar to the .PRINT statement, with the parameter *<analysis type>* being DC, AC, noise, or transient analysis. The parameter *<output variables>* can be a series of variables such as node voltages and component currents. As an example of the use of the .PLOT statement, consider

.PLOT AC V(5,0)

This statement causes the voltage between nodes 5 and ground of an AC analysis to be stored as a plot in the circuit output file. Although plots created using the .PLOT statement are useful, they can be difficult to interpret, since characters such as "+" and "*" are used for plotting. The .PLOT statement has been superseded by a graphics program called *Probe*[1].

2.5.3 *Graphics outputs — Probe*

PSpice allows the user to view the simulation results graphically with much higher resolution. This is achieved using the program *Probe*. This program is separate from PSpice, but it is part of the PSpice simulation package. To view the simulation results using *Probe*, the user must first specify the statement:

.PROBE [output variables]

This statement instructs PSpice to write the simulation results to a data file called PROBE.DAT. This file is later used by the program *Probe* to produce the simulation results in graphic format. Note that the parameter *[output variables]* can be node voltages and/or component currents. If no *[output variables]* are specified, *Probe* will save *all* the node voltages and component currents of the circuit that are then ready for plotting. The allowed output variables in the .PROBE statement are shown in Tables 2.8 and 2.9. Table 2.8 shows the output variables available for various analyses. For AC analysis, the output variables listed in the general form of Table 2.8 can be used, but a suffix can also be added to them appropriate to the measurement required. Table 2.9 shows the available suffixes. The program *Probe* can also be used to

Table 2.8   DC, AC, and Transient Analyses Output Variables

General form	Meaning	Example
V(<*node*>)	Voltage at specified node	V(3) Voltage between nodes 3 and 0
V(<+ *node*>, <– *node*>)	Voltage difference between + and – nodes	V(2,3) Voltage between nodes 2 and 3
V(<*name*>)	Voltage across a two-terminal component; the <*name*> of such a component is given in Table 2.10	V(R1) Voltage across R1
Vxy(<*name*>)	Voltage at xy, where x and y are nodes of a three- or four-terminal component; the <*name*> of such a component is given in Table 2.11	VBE(Q1) Base-emitter voltage of Q1
Vz(<*name*>)	Voltage at one end of a transmission line	VA(T1) Voltage at port A of transmission line T1
I(<*name*>)	Current through a two-terminal component; the <*name*> of such a component is given in Table 2.10	I(D1) Current through diode D1
Ix(<*name*>)	Current into node x, where x is a node of a three- or four-terminal component; the <*name*> of such a component is given in Table 2.11	IE(Q2) Emitter current of Q2

Table 2.9   AC Analysis Output Variable Suffixes

Suffix	Meaning	Example
None	Magnitude	V(2)
M	Magnitude	VM(2)
dB	Magnitude (dB)	VdB(2)
P	Phase	VP(2)
G	Group delay	VG(2)
R	Real part	VR(2)
I	Imaginary part	VI(2)

perform various arithmetic functions on the simulation output variables, ranging from simple operations (+,-,*,/) to complex differentiation and integration. The aim here is to give an introduction to *Probe*; to appreciate the full capabilities of the program, the reader is directed to Reference #1.

Table 2.10 Two-Terminal Components

<name>	Meaning
C	Capacitor
D	Diode
E	Voltage-controlled voltage source
F	Current-controlled current source
G	Voltage-controlled current source
H	Current-controlled voltage source
I	Independent current source
L	Inductor
R	Resistor
S	Voltage-controlled switch
V	Independent voltage source
W	Current controlled switch

Table 2.11 Three- and Four-Terminal Components

<name>	Node Abbreviations
B (GaAsFET)	D (Drain), G (gate), S (source)
J (Junction FET)	D (Drain), G (gate), S (source)
M (MOSFET)	D (Drain), G (gate), S (source), B (bulk, substrate)
Q (Bipolar transistor)	C (Collector), B (base), E (emitter), S (substrate)

2.6 Running PSpice — the control shell

The circuit description, in terms of components, input signal(s), analysis type(s), and output options including the .PROBE statement, is usually referred to as the circuit input file or the circuit netlist. A circuit input file is often created using a text editor. This represents the first step in the simulation process, as discussed at the beginning of this chapter. Having created the input file, the next step is to run PSpice and obtain the simulation results.

PSpice automates the above simulation steps using the *Control Shell* software[1]. The *Control Shell* consists of an editor, which can be used for creating input files, and a "shell" program, from which PSpice and *Probe* can be run. The *Control Shell* is initiated from the DOS system prompt by typing the command

PS

Figure 2.8 shows the start-up screen for the *Control Shell*. For example, "Files" allows an input file to be created, "Analysis" is used to run PSpice, and "Probe" provides the graphic outputs.

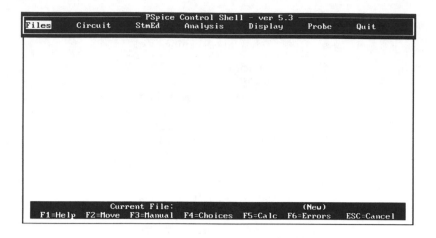

Figure 2.8   Start-up screen of the *Control Shell* program.

It is possible to use other text editors to create PSpice input files, but some text editors add undesired characters that may cause PSpice to report error messages.

2.7   Simulation examples

It is now possible to use all the above commands to input circuits, simulate their performance, and output the results. Two examples are considered, with the first based on a simple NPN transistor. This example is chosen to demonstrate the use of DC sweep analysis. The second example, based on a transistor amplifier circuit, is intended to show how to use PSpice to obtain the DC biasing points, and the frequency response of the amplifier circuit.

Example 2.7.1 NPN Transistor Circuit
Here, PSpice will be used to generate the output characteristic curve of an NPN transistor in a common emitter configuration. Figure 2.9 shows the circuit with the assigned nodes.

To create the input file of this circuit, first load *Control Shell* by typing "PS", as discussed in Section 2.6, and Figure 2.8 will then appear. Select the option "Files" from the menu, and then the command "Current File" from the submenu. A name must be given to the input file to enter the text edit mode, and in this case, let us assume it is "EX271.CIR". The input file name must have the extension ".CIR", which *Control Shell* adds automatically. Now select the command "Edit" from the "Files" submenu, and the circuit is ready for entry to PSpice using the various description statements as shown in Figure 2.10.

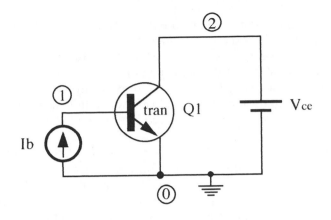

Figure 2.9 Circuit of Example 2.7.1.

This shows that a current source, Ib, is used to describe the base current, while a voltage source, Vce, is used to describe the collector-emitter voltage. The .DC statement causes Vce to be increased from 0 to 5V in steps of 0.1V, while Ib is increased from 0 to 60μA in steps of 10μA. This statement instructs PSpice to perform "nested" DC sweep analysis, with Vce being the first sweep and Ib the second sweep. The entire Vce sweep will be done for each value of Ib. The transistor nodes are described using the Q1 statement, and the transistor model name, chosen arbitrarily, has been assumed to be "tran". The .MODEL statement specifies a NPN transistor with default model parameters. The .PROBE statement instructs PSpice to save the result of all the node voltages and currents of the circuit.

```
┌──────────────── PSpice Control Shell - ver 5.3 ─────────────────┐
│══════════════ Circuit Editor   line:  1 col:  1    [Insert] ═══ │
│NPN Transistor                    ; title line                   │
│* Circuit description             ; comment line                 │
│Ib 0 1                            ; current source               │
│Vce 2 0                           ; voltage source               │
│*                                                                │
│.DC Vce 0V 5V 0.1V Ib 0u 60u 10u  ; nested DC sweep of Vce & Ib  │
│*                                                                │
│Q1 2 1 0 tran                     ; transistor connection & model name │
│.MODEL tran NPN                   ; NPN with default model parameters │
│*                                                                │
│.PROBE                            ; graphic output               │
│.END                              ; end of circuit input file    │
│                                                                 │
│                                                                 │
│                                                                 │
│          Current File: A:EX271.CIR            Loaded            │
│  F1=Help  F2=Move  F3=Manual  F4=Choices  F5=Calc  F6=Errors  ESC=Cancel │
└─────────────────────────────────────────────────────────────────┘
```

Figure 2.10 PSpice input file of Example 2.7.1.

The first line of the input file must be a title and the last line an .END statement. To improve the clarity of the input file, PSpice allows the user to include comments lines in the file. A comment line is included by putting an asterisk "*" at the beginning of the line, as shown in Figure 2.10. (This provides a good way of removing a component from an input file during circuit testing.) PSpice also allows comments to be added to description statements. This is achieved by starting the comment with a semicolon ";" (see Figure 2.10).

To run PSpice, select "Analysis" from the *Control Shell* menu and then the command "Run PSpice" from the submenu. If errors are encountered in the input file, PSpice will terminate, and a list of errors will be produced in the circuit .OUT file. This file can be viewed by selecting the command "Browse Output" from the "Files" submenu. If there are no errors, then PSpice will perform the simulation and return to the *Control Shell* menu. To plot the simulation results, *"Probe"* should be selected from the main menu. *Probe* usually plots the sweep variable of the analysis as the x-axis (current or voltage for DC sweep, frequency for AC analysis, and time for transient analysis). The y-axis can be any node voltages and current of the circuit. This depends on the output variables specified in the .PROBE statement. In this example, no output variables were specified (Figure 2.10), so all circuit node voltages and currents will be available. To obtain a graph of the transistor output characteristics (Ic against Vce for a given Ib), first select "Add_trace" from the *Probe* menu, then type "Ic(Q1)". The variable Ic(Q1) means the collector current through transistor (Q1) (see Table 2.8).

Figure 2.11 shows the PSpice simulation of the transistor output characteristics which reproduces the familiar transistor curves. Note the y-axis variable name appears just below the origin.

Example 2.7.2 Amplifier Circuit
Consider the common emitter transistor amplifier shown in Figure 2.12. In this example, PSpice will be used to simulate

1. Transistor operating points
2. Amplifier frequency response (gain and phase) over the frequency range of 1Hz to 30kHz (assume that the transistor has a current gain of 50)

The PSpice input file of the amplifier circuit is given in Figure 2.13. This file shows that the power supply has been described by voltage source Vcc and that the component values have been expressed using the scale suffix notation (see Table 2.2). The model name (tran) specifies a NPN transistor having default model parameters, apart from the current gain BF = 50 (the default value of BF = 100).

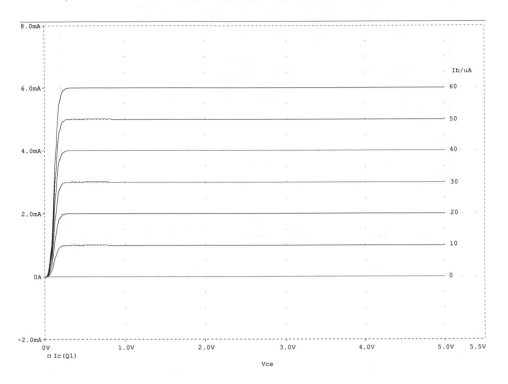

Figure 2.11 NPN transistor output characteristics.

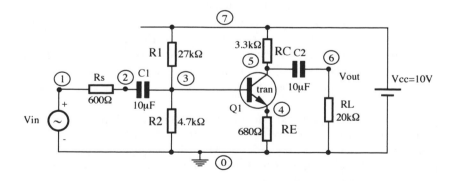

Figure 2.12 Circuit of Example 2.7.2.

```
┌─────────────────────── PSpice Control Shell - ver 5.3 ───────────────────────┐
│═════════════════ Circuit Editor    line:   1 col:   1      [Insert] ═════════│
│Common emitter transistor circuit                 ; title line                │
│*                                                                             │
│Vin 1 0 AC 1                         ; ac input signal of 1V                   │
│.AC OCT 100 1 30kHz                  ; frequency analysis range               │
│Vcc 7 0 DC 10V                       ; 10V power supply                        │
│*                                                                             │
│RS 1 2 600                           ; RS between nodes 1 & 2, value 600 Ohm   │
│C1 2 3 10uF                                                                   │
│R1 3 7 27K                                                                    │
│R2 3 0 4.7K                                                                   │
│RC 5 7 3.3K                                                                   │
│RE 4 0 680                                                                    │
│C2 5 6 10uF                          ; C2 between nodes 5 & 6, value 10uF      │
│RL 6 0 20K                                                                    │
│Q1 5 3 4 tran                        ; transistor connections & model name    │
│.MODEL tran NPN [BF=50]              ; NPN transistor with current gain=50     │
│.OP                                  ; circuit operating points info command  │
│.PROBE                               ; graphic output                         │
│.END                                 ; end of circuit input file              │
│                                                                             │
│              Current File: A:EX272.CIR                 Loaded                │
│   F1=Help  F2=Move  F3=Manual  F4=Choices  F5=Calc  F6=Errors    ESC=Cancel  │
└─────────────────────────────────────────────────────────────────────────────┘
```

Figure 2.13   PSpice input file of Example 2.7.2.

To discover the biasing points and the various currents of the amplifier transistor, the .OP statement is required, as shown in the Figure 2.13. This statement outputs directly into the .OUT file. The .OP output is included in the circuit output file under the heading "operating point information", as shown in Box 2.1. Also included in this information are some of the transistor small signal parameters. As well as the operating point information, the output file usually contains the circuit description and a list of the circuit node voltages. This list of node voltages is usually referred to in the circuit output file as the "small signal bias solution", and PSpice calculates it by default.

The AC independent source description statement (i.e., the Vin statement, Figure 2.13) together with the .AC statement cause PSpice to calculate the frequency response of the amplifier over the specified frequency range of 1Hz to 30kHz with 100 frequency points on a logarithmic (octave) scale. Note that an AC input signal source must be specified, and in this case it is Vin with 1V amplitude and 0° phase shift.

The simulated gain frequency response of the amplifier is shown in Figure 2.14. This figure is obtained by typing "V(6)", having selected "Add_trace" from the *Probe* menu. The phase response is shown in Figure 2.15. This is obtained by typing "VP(6)", once the graph of V(6) has been deleted from the screen. The reader is referred to Table 2.8 for

Box 2.1 PSpice Output File of Example 2.7.2

```
Common emitter transistor circuit ; title line
*   CIRCUIT DESCRIPTION
Vin 1 0 AC 1                        ; AC input signal of 1V
.AC OCT 100 1 30K                   ; frequency analysis range
Vcc 7 0 DC 10V                      ; 10V power supply
*
RS 1 2 600                          ; RS between nodes 1 & 2, value 600 ohm
C1 2 3 10uF
R1 3 7 27k
R2 3 0 4.7k
RC 5 7 3.3k
RE 4 0 680
C2 5 6 10uF                         ; C2 between nodes 5 & 6, value 10uF
RL 6 0 20k
Q1 5 3 4 tran                       ; transistor connections & model name
.MODEL tran NPN [BF = 50]           ; NPN transistor with current gain = 50
.OP                                 ; circuit operating points info command
.PROBE                              ; graphic output
.END                                ; end of circuit input file
*********************************************************************************
********    BJT MODEL PARAMETERS
*********************************************************************************
                    tran
                    NPN
           IS 100.000E-18
           BF 50
           NF 1
           BR 1
           NR 1
****SMALL SIGNAL BIAS SOLUTION  TEMPERATURE = 27.000 DEG C
*********************************************************************************
***

NODE VOLTAGE NODE VOLTAGE NODE VOLTAGE NODE VOLTAGE
  (1)    0.000    (2)    0.000    (3)    1.4436    (4)    0.6700

  (5)    6.6806    (6)    0.000    (7)    10.000

       VOLTAGE SOURCE        CURRENTS
       NAME                  CURRENT
       Vcc                   −1.292E-03

       TOTAL POWER DISSIPATION        1.29E-02 WATTS
```

Box 2.1 (continued)   PSpice Output File of Example 2.7.2

```
****OPERATING POINT INFORMATION  TEMPERATURE = 27 DEG C
*********************************************************************************
***
*** BIPOLAR JUNCTION TRANSISTORS

NAME            Q1
MODEL           tran
IB              9.76E-06
IC              9.76E-04
VBE             7.74E-01
VBC             -5.34E+00
VCE             6.11E+00
BETADC          50
GM              3.77E-02
RPI             2.65E+03
RX              0.00E+00
RO              1.00E+12
CBE             0.00E+00
CBC             0.00E+00
CBX             0.00E+00
CJS             0.00E+00
BETAAC          50
FT              6.00E+17
```

the other options available to plot output variables of AC analysis. It should be noted that the simulated results agree well with the theoretical results (which are based on analytical expressions) given in Reference 2.

If it is desired to obtain the amplifier frequency response in the form of a table, then the following statement:

.PRINT AC V(6)

must be specified in the input file. This statement causes PSpice to print in the circuit output file a table of two columns, with the first being the frequency and the second the amplifier output voltage at node 6. The frequency range is determined by the .AC statement.

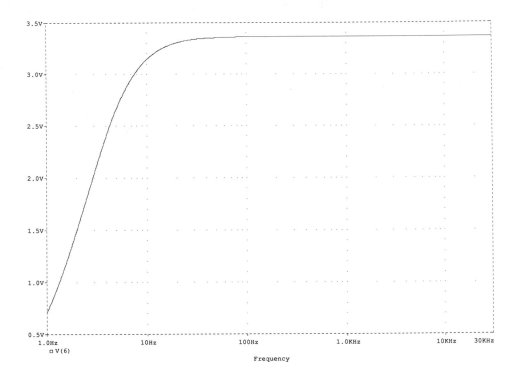

Figure 2.14   Frequency response (gain) of Example 2.7.2.

2.8   Schematic capture

So far, circuit descriptions have been performed using components, independent sources, analysis type(s), and .PROBE statements. These statements form the circuit input file or the circuit netlist. Electronic circuits are increasing in complexity all the time; this means that describing or modifying a complex circuit using component statements can easily prove to be a time-consuming and error-prone task. Clearly one method of simplifying this process is to draw the circuit and generate the netlist automatically. Recently PSpice introduced the program *Schematics*[3], which performs just that. This program is available only for the Sun workstations and DOS/Windows PSpice versions. Although this book is based on PSpice DOS version, the examples are equally valid for other PSpice versions, including the DOS/Windows.

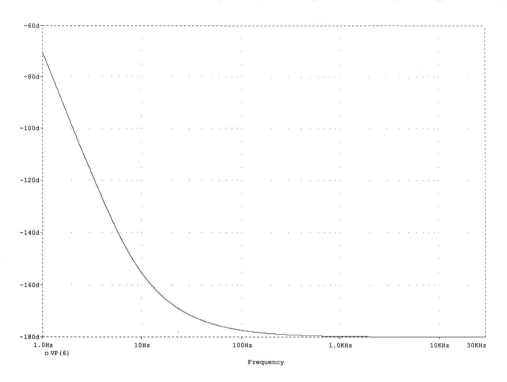

Figure 2.15   Frequency response (phase) of Example 2.7.2.

Schematics is a menu-driven program and is based on using part symbols picked from the PSpice library to represent components, and using signal generators and lines to represent wire connections. Once the circuit drawing is finished, the program will generate the circuit netlist automatically. The PSpice user's guide to schematics capture deals with this topic[3].

Irrespective of whether the circuit netlist is generated manually (i.e., description statements) or using schematics capture programs, it is important to understand the basic rules for producing circuit netlists or input files. This enables the user to utilise the full capabilities of PSpice.

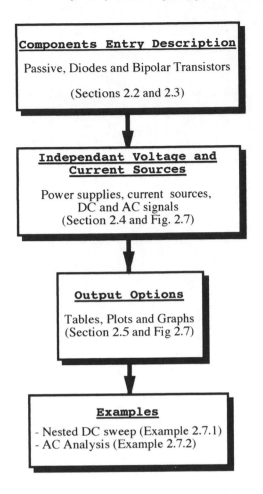

Figure 2.16 Flow chart of Chapter 2 summary.

2.9 Chapter summary

The best summary of this chapter is a flow chart of the whole simulation process, as shown in Figure 2.16.

References

1. MicroSim Corporation, *Circuit Analysis — User's Guide Manual* (The Design Center), Version 5.3, Irvine, CA, January 1993.
2. Bogart, F. B., *Electronics Devices and Circuits*, 2nd edition, Merrill Publishing Company, Columbus, OH, 1990, pp. 205-206.
3. MicroSim Corporation, *Schematic Capture — User's Guide Manual* (The Design Center), Version 5.2, Irvine, CA, July 1992.

chapter three

Transient and fourier analyses

The previous chapter has shown how DC and AC circuit analyses are performed. This chapter discusses the simulation of transient response and Fourier analysis. Examples of passive and active circuits will be given to illustrate the various analyses.

3.1   Transient independent sources

These sources are used when the response of a circuit to a transient signal is required. Two statements are required to perform transient analysis: the first is an independent source description statement, and the second is a .TRAN statement, which sets the period of analysis. The basic form of an independent source description statement is

<source name> <+ node> <- node> <type> <value>

where the parameter *<source name>* is the independent voltage or current source. These sources are identified by PSpice as components starting with the letters V and I, respectively, as discussed in Section 2.4 and Table 2.6 of Chapter 2. The parameters *<+ node>* and *<- node>* show how the source is connected within the circuit. There are five types, parameter <type>, of time-dependent sources available in PSpice. These are exponential (EXP), pulse (PULSE), sinusoidal (SIN), single-frequency frequency-modulated (SFFM), and piece-wise linear (PWL) sources.

3.1.1   Exponential independent source (EXP)

This source defines a voltage or current with exponential rise time and exponential fall time. Thus, the description statement of an exponential independent voltage source is

<Vxxx> <+ node> <- node> EXP (v1 v2 td1 tau1 td2 tau2)

where the parameter *<Vxxx>* is the voltage source name, and *EXP* specifies an exponential transient source. The exponential signal

Table 3.1   Exponential Signal Parameters

Parameter	Meaning	Default value	Unit
v1	Initial value	None	V
v2	Final value	None	V
td1	Rise delay time	0.0	s
tau1	Rise time constant	*<print interval>* (see Section 3.2)	s
td2	Fall time delay	*td1+ print interval* (see Section 3.2)	s
tau2	fall time constant	*<print interval>* (see Section 3.2)	s

parameters are explained in Table 3.1. Signal parameters that have no default values must be specified by the user; examples are the initial and final values (*v1* and *v2*) of the signal.

As an example, the general form of the voltage exponential source

<Vxxx> <+ node> <– node> EXP (v1 v2 td1 tau1 td2 tau2)

becomes

Vin 1 0 EXP (1V 2V 0.4us 1us 5us 0.2us)

This statement describes a voltage source (Vin), connected between nodes 1 and 0, which has an exponential form described by the parameters shown in Figure 3.1.

It should be noted that for every voltage source discussed here, there is a corresponding current source.

3.1.2   Pulse independent source (PULSE)

This source defines a voltage or current with pulse characteristics. This can be typically square or triangular waveforms. The description statement of a pulse voltage source is

<Vxxx> <+ node> <– node> PULSE (v1 v2 td tr tf pw per)

where the parameter *PULSE* specifies a pulse independent source. The pulse signal parameters are explained in Table 3.2.

As an example, the general form of a voltage pulse source

<Vxxx> <+ node> <– node> PULSE (v1 v2 td tr tf pw per)

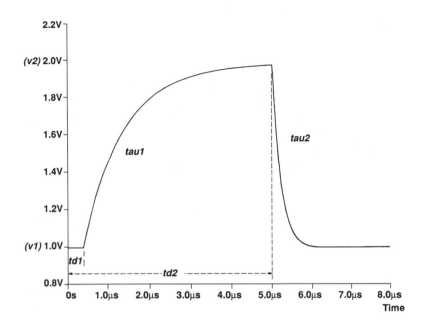

Figure 3.1 PSpice description of an exponential waveform.

becomes

Vin 1 0 PULSE (–5V 5V 0.2us 0.1us 0.1us 0.4us 1us)

This statement describes a voltage source (Vin) connected between nodes 1 and 0. This source produces a 1MHz square waveform (or 1us period) as shown in Figure 3.2. The waveform repeats itself except for the initial delay time of 0.2us.

Table 3.2 Pulse Signal Parameters

Parameter	Meaning	Default value	Unit
v1	Initial value	None	V
v2	Final value	None	V
td	Time delay	0.0	s
tr	Rise time	*<print interval>* (see Section 3.2)	s
tf	Fall time	*<print interval>* (see Section 3.2)	s
pw	Pulse width	*<final time>* (see Section 3.2)	s
per	Period	*<final time >* (see Section 3.2)	s

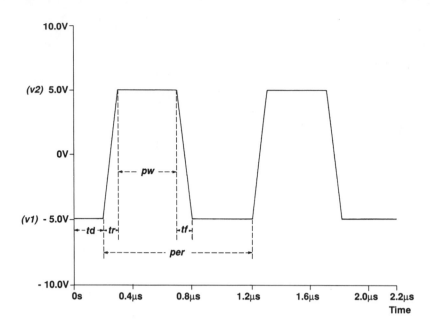

Figure 3.2   PSpice description of pulse waveform.

To generate a 100Hz triangular waveform with amplitude of –2 to +2V, the following statement is required:

Vin 1 0 PULSE (–2V 2V 0s 5ms 5ms 1ps 10ms)

The pulse width (*pw*) has been set to a very small value, 1ps, or 1E-12s. The simulated triangular waveform is shown in Figure 3.3. It has been assumed that the input signal source is connected between nodes 1 and 0.

3.1.3   Sinusoidal independent source (SIN)

This source defines a sinusoidal voltage or current. The statement of sinusoidal voltage source is

<Vxxx> <+ node> <– node> SIN (vo va freq td df phase)

where the parameter *SIN* specifies a sinusoidal independent source. The signal parameters are explained in Table 3.3. The voltage of the sinusoidal signal starts at <*vo*> and stays there for <*td*> seconds. Then, the voltage becomes an exponentially damped sine wave. This damped sine wave is defined by the equation[1]

$$vo+va*\sin(2\pi*(freq*(TIME-td)+phase/360))*e^{-(TIME-td)*df}$$

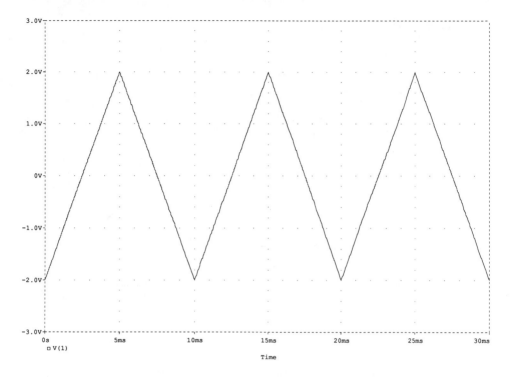

Figure 3.3   Simulated triangular waveform (amplitude = 4V, period = 10ms).

where TIME is the PSpice internal sweep variable used in the transient analysis.

If the parameters *<td>*, *<df>*, and *<phase>* are not specified by the user (i.e., set to their default values), the SIN independent source produces a sine waveform. As an example, the statement

Vin 1 0 SIN (0V 1V 1kHz)

produces the waveform shown in Figure 3.4. It should be noted that the SIN waveform is for transient analysis only. It does not have any effect

Table 3.3   Sinusoidal Signal Parameters

Parameter	Meaning	Default value	Unit
vo	Offset voltage	None	V
va	Peak amplitude	None	V
freq	Frequency	1 / *<final time>* (see Section 3.2)	Hz
td	Delay time	0.0	s
df	Damping factor	0.0	1/s
phase	Phase	0.0	degrees

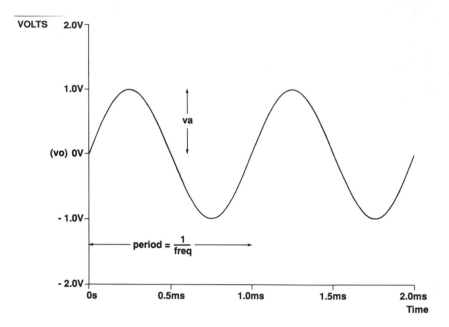

Figure 3.4   PSpice description of a sinusoidal waveform.

during AC analysis. To perform AC analysis, see Section 2.4.2 and Example 2.7.2 of Chapter 2.

3.1.4   *Single-frequency frequency-modulated source (SFFM)*

This source generates a voltage or current single-frequency frequency-modulated (SFFM) signal. The description statement of a SFFM voltage source is

<Vxxx> <+ node> <– node> SFFM (vo va fc mdi fs)

where the parameter *SFFM* defines a single-frequency frequency-modulated source. The parameters of this signal are explained in Table 3.4. The modulation index (*mdi*) is defined as the deviation (the peak variation in carrier frequency in Hertz) divided by the signal frequency. The SFFM voltage signal is defined by the equation[1]

vo+va*sin(2πfc*TIME+mdi*sin(2πfs*TIME))

where TIME is the PSpice internal sweep variable used in the transient analysis. Figure 3.5 shows an SFFM waveform, which is described by the following statement:

Vin 1 0 SFFM (0V 1V 10kHz 5 1kHz)

Table 3.4 SFFM Signal Parameters

Parameter	Meaning	Default value	Unit
vo	Signal offset	None	V
va	Signal amplitude	None	V
fc	Carrier frequency	1 / *<final time>* (see Section 3.2)	Hz
mdi	Modulation index	None	None
fs	Signal frequency	1 / *<final time>* (see Section 3.2)	Hz

This signal has a zero offset voltage, 1V peak amplitude, 5kHz of deviation from a carrier signal of 10kHz. Note that in this example, the values of the carrier frequency (*fc*) and the modulation index (*mdi*) were chosen arbitrarily for illustration purposes. In practice, a higher carrier frequency and a lower modulation index are normally used.

3.1.5 Piece-wise linear independent source (PWL)

Any arbitrary voltage or current waveform may be described using a piece-wise linear (PWL) independent source. Each pair of time-voltage (or time-current) values represents the coordinate of a point on the

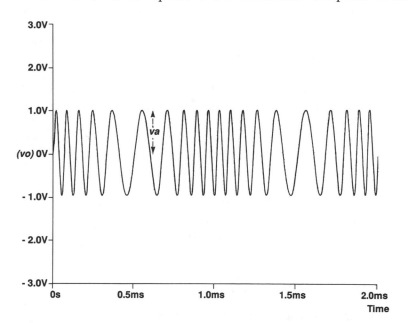

Figure 3.5 SFFM waveform (signal frequency = 1kHz, carrier frequency = 10kHz, modulation index = 5).

Table 3.5   PWL Signal Parameters

Parameter	Meaning	Default Value	Unit
tn	nth-Time value	None	s
vn	nth-Voltage value	None	s

waveform. Up to 3995 pairs may be used. The voltage (or current) at times between the intermediate points is determined by PSpice using linear interpolation. The description statement of a PWL voltage source is

<Vxxx> <+ node> <– node> PWL + ((t1,v1) (t2,v2) (t3,v3)...(tn,vn))

Note that this statement may well be longer than one line. In this case, PSpice allows the statement to be extended to a second line by use of the "+" sign. The parameter PWL specifies a piece-wise linear source. The parameters of the PWL signal are explained in Table 3.5.

As an example, the statement

Vin 1 0 PWL ((0us,5V) (2us,5V) (2.0001us,0V) (3us,0V)
+ (3.0001us,-5V) (4us,-5V) (4.0001us,0V) (4.5us,0V)
+ (4.50001us,5V) (5us,5V) (5.0001us,0V) (7us,0V))

represents a voltage independent source connected between nodes 1 and 0 with a PWL description of the signal shown in Figure 3.6.

PSpice requires that the time value coordinates increase from one coordinate to the next. Hence, when describing an "instantaneous" signal change the same time value cannot be used. In the given example, voltages at 2.0 and 2.0001µs represent the change in voltage from +5 to 0V in 2µs. The "+" sign in the above description signifies continuation of the statement.

Clearly, it can be seen from this example that in order to describe complex waveforms using PWL independent voltage or current sources, a large number of statement lines will be needed. To simplify this, PSpice semiautomates the process using the program *Stimulus Editor* (*StmED*). Additionally, this program can be used to generate any of the independent voltage or current sources already discussed. Appendix B shows how the program *StmEd* is used to describe independent voltage or current waveforms.

3.2   The .TRAN statement

This statement is used to specify the period of time over which the circuit transient analysis is to be performed. The .TRAN statement together with the required transient independent source description

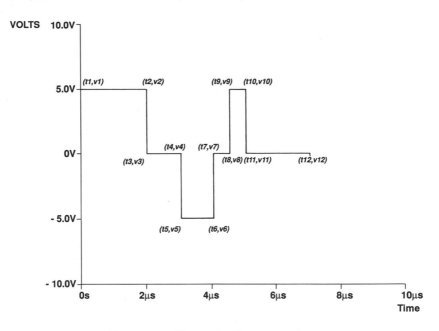

Figure 3.6 Piece-wise linear waveform.

statement (i.e., EXP, PULSE, ...) allow the user to perform transient analysis. The basic form of the .TRAN statement is

.TRAN <print interval> <final time> [no print interval] + [time step] [UIC]

This statement is one of the few PSpice statements that is not immediately obvious in its application, and it is best illustrated by the use of an example. Consider the waveform shown in Figure 3.7.

This waveform has been generated by the statements

Vin 1 0 SIN (0V 1V 1kHz)
.TRAN 0.01ms 2ms 0ms 0.01ms

The first statement describes a 1V, 1kHz sine wave. The .TRAN statement is explained as follows:

> *<print interval>* determines the number of print points across the span. In the example given, 0.01ms means 200 points across the printout. Note that if the print interval chosen is too coarse, erroneous curves may result. Figure 3.8 shows the same waveform with a print interval of 0.1ms.
> *<final time>* is the span of the analysis, and in this example it is 2ms. Note that PSpice always starts transient analysis at time $t = 0$.

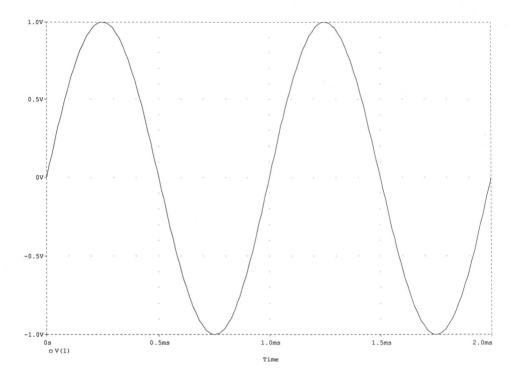

Figure 3.7   1kHz sine wave with 200 print points.

In addition to the compulsory parameters (*<print interval>* and *<final time>*), the following optional parameters can be specified:

[*no print interval*] — although PSpice starts the analysis at time t = 0, it is possible to suppress an early part of the simulation result when printed or plotted. In the example given, [no print interval] has been set to 0ms (the default value). Figure 3.9 shows the original waveform with [*no print interval*] set to 0.4ms.

[*time step*] is the maximum period between individual calculations PSpice uses when carrying out the transient simulation. If not specified, PSpice uses the default value of *<final time/50>*, which would have been 0.04ms for the example used if it had not been specified as 0.01ms. There is a compromise when deciding on the value of the [*time step*] parameter for a simulation. A small time step will ensure accuracy at the expense of longer simulation time. A large time step will reduce simulation accuracy, and signal transient features may be distorted or even omitted. It is of interest to note that PSpice does not necessarily use equally spaced calculation points.

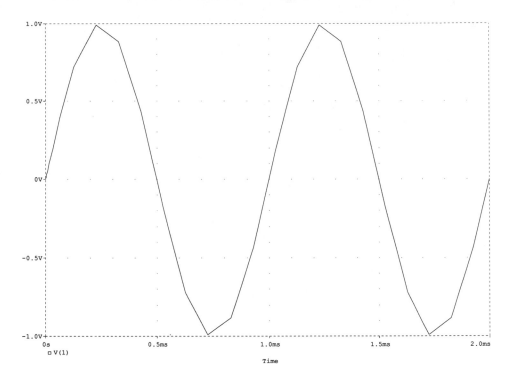

Figure 3.8   1kHz sine wave with 20 print points.

[*UIC*] stands for "use initial condition" and is used when some
initial conditions on components are required to be specified.
The default value of the parameter [*UIC*] is zero. Capacitors can
have initial voltages, and inductors can have initial currents.

These initial conditions can be specified as part of the component
description statement. As an example, consider the statements

.TRAN 1ms 10ms UIC
C1 2 3 0.1uF IC = 0.3V
L1 3 0 1mH IC = 0.1mA

The first statement includes the [*UIC*] command. The second statement
shows the capacitor (C1) connected between nodes 2 and 3 with an
initial voltage of 0.3V. The third statement shows that the inductor (L1)
is connected between nodes 3 and ground with an initial current of
0.1mA.

There are other methods that can be used to set simulation initial
conditions. These methods are discussed in Appendix C, which covers
the simulation of oscillators.

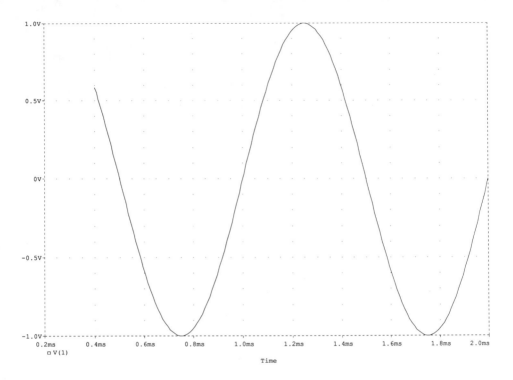

Figure 3.9   1kHz sine wave with 200 print points starting at 0.4ms.

3.3   Simulation examples

To illustrate the use of transient analysis, two examples are considered. The first example is based on a push-pull amplifier circuit, while the second is based on a lowpass filter. Further examples of transient analysis are given in Appendices C and D.

Example 3.3.1 Push-Pull Amplifier
Figure 3.10 shows a push-pull amplifier circuit.

The PSpice input file of this amplifier is given in Listing 3.1. A sine waveform of 5V and 1kHz has been used as the input signal. The .TRAN statement instructs PSpice to simulate the amplifier output over 2ms with a time step of 0.02ms. Figure 3.11 shows the simulated input and output signals of the amplifier circuit.

Clearly the crossover distortion[2] can be seen, as well as the reduction in the output signal amplitude, which is due to the base-emitter voltage drop across each transistor.

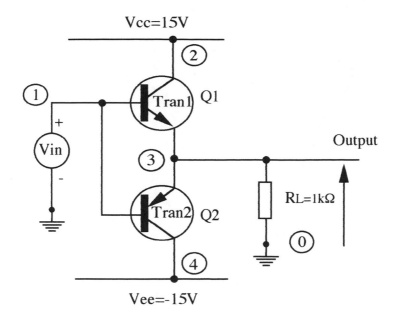

Figure 3.10   Circuit of Example 3.3.1.

Listing 3.1   Input File of Example 3.3.1

```
Complementary push-pull amplifier (Figure 3.10) ;   title line
*
Vin 1 0 SIN (0 5V 1kHz)      ; 5V, 1kHz sine input signal
.TRAN 0.02ms 2ms 0ms 0.02ms ; transient analysis range
*
Vcc 2 0 15V                 ;
Vee 4 0 -15V                ; power supplies
*
Q1 2 1 3 tran1
Q2 4 1 3 tran2              ; tran connections and model  name
.MODEL tran1 NPN
.MODEL tran2 PNP           ; default tran model parameters
*
RL 3 0 1k
*
.PROBE V(1), V(3)          ; graphic outputs of input & output
*
.END                       ; end of circuit input file
```

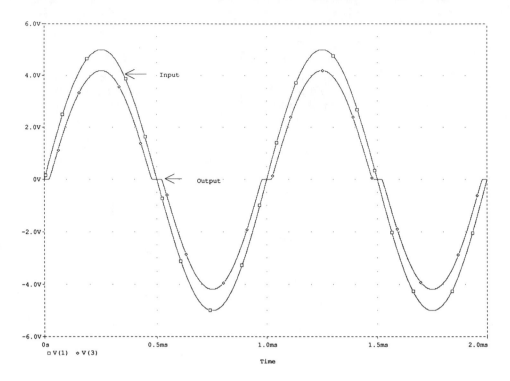

Figure 3.11   Input and output waveforms of Example 3.3.1.

Example 3.3.2 Lowpass Passive Filter

Consider a requirement for a lowpass filter with the following specification:

> Passband edge = 1MHz
> Stopband edge = 1.5MHz
> Stopband attenuation ≥30dB
> Source and load impedance = 75Ω

This filter specification can be met using a fifth-order elliptic circuit[3] as shown in Figure 3.12.

It is necessary to examine the transient filter response when it is fed by a pulse of 2V peak-to-peak amplitude, with 10ns rise-and-fall time and 8μs period. It is also necessary to simulate the filter frequency response (attenuation and phase) over the frequency range of DC-5MHz. The input file of the filter is given in Listing 3.2. PSpice allows the specification of multiple analyses in one statement, as shown in Listing 3.2 (line 5). To specify the frequency and the transient range over which each analysis is performed, .AC and .TRAN statements are

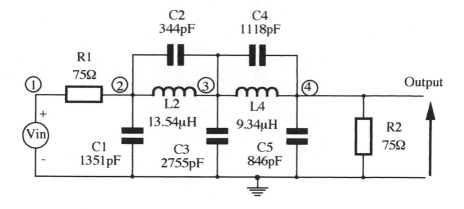

Figure 3.12   Circuit of Example 3.3.2.

Listing 3.2   Input File of Example 3.3.2

```
1MHz Elliptic lowpass LC filter (Figure 3.12); title line
*
* AC and transient input signals            ; comment line
*
Vin 1 0 AC 1V PULSE (-1V 1V 0s 10ns 10ns 4us 8us)
*
.AC LIN 500 0.01Hz 5meg        ; frequency analysis range
*
.TRAN 0.03us 60us 0u 0.03us    ; transient analysis range
*
* Filter Components
R1 1 2 75
C1 2 0 1351pF
C2 2 3 344pF
L2 2 3 13.54uH
C3 3 0 2755pF
C4 3 4 1118pF
L4 3 4 9.34uH
C5 4 0 846pF
R2 4 0 75
*
.PROBE V(1), V(4)        ; graphics outputs of V(1) & V(4)
.END
```

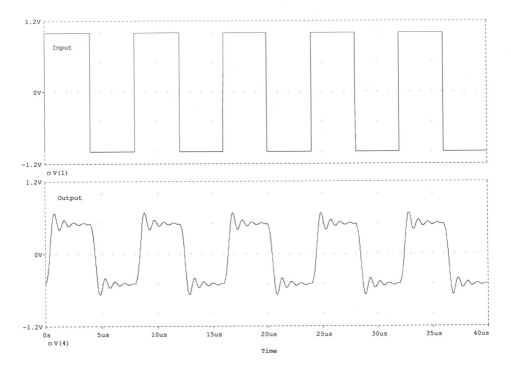

Figure 3.13   Transient response of Example 3.3.2.

required (see the listing). PSpice does not allow frequency response analysis to start at DC; therefore, in this example the analysis has been specified to start at 0.01Hz. Figure 3.13 shows the simulated transient response of the filter, which shows that the output signal has a considerable overshoot or ringing. The reduction in the output signal amplitude is due to the filter being equally terminated (i.e., R1 = R2 = 75Ω). To divide the screen for various plots as in Figure 3.13, select the option "Plot_control" from the *Probe* main menu, and then "Add_plot" from the submenu, followed by "Exit" to go back to the main menu to start plotting.

Figure 3.14 shows the simulated frequency response of the filter. As this is an equally terminated filter (i.e., R1 = R2) a 6dB loss is incurred at DC.

Probe allows the user to determine the values of various points on a graph or on a waveform using two cursors, called C1 and C2. For example, the filter frequency response (Figure 3.14) shows that the attenuation at DC (489.386f) is –6.02dB, while at 1.5MHz it is –39.51dB. *Probe* also provides the difference between the two cursors, C1 and C2, and in this case the attenuation at 1.5MHz with respect to DC is –33.49dB. The cursor option is selected by choosing "Cursor" from the *Probe*

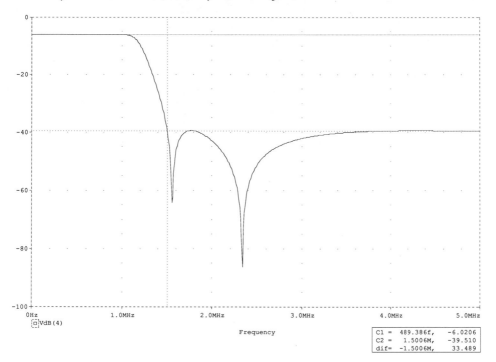

Figure 3.14   Frequency response (attenuation) of Example 3.3.2.

menu. The trace for the first cursor is changed by pressing the <Ctrl ← > and <Ctrl → > key combinations. The trace for the second cursor is changed by using the key combinations <Shift Ctrl ← > and <Shift Ctrl → >.

Figure 3.15 shows the simulated phase response of the filter. If it is necessary to plot the filter group delay, which is the derivate of phase with respect to frequency (–dPHASE/dFREQUENCY), then type "Vg(4)" in place of "Vp(4)", as shown at the bottom of Figure 3.15.

3.4   Fourier analysis (text mode)

Fourier's theorem states that all waveforms can be constructed by a suitable addition of a series of sine waves. Any periodic signal can therefore be represented by an equation. If $f(t)$ is the periodic signal, the equation that models it is

$$f(t) = a_0 + a_1\sin(\omega_0 t + \phi_1) + a_2\sin(2\omega_0 t + \phi_2) + a_3\sin(3\omega_0 t + \phi_3) + \dots$$

where the term a_o is the DC component of the signal. The terms $a_1, a_2, a_3, \dots$, represent the amplitude of the Fourier components or harmonics. The terms ϕ_1, ϕ_2, ϕ_3 represent the values of the phases of the Fourier components. The term ω_o is the fundamental frequency of the signal.

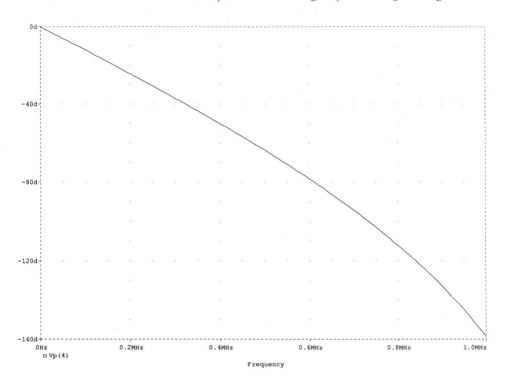

Figure 3.15 Frequency response (phase) of Example 3.3.2.

PSpice allows the user to perform Fourier analysis using the .FOUR statement. This statement calculates the signal Fourier components, including the DC, fundamental, and second to ninth harmonics. The amplitude and phase of these components are then tabulated in the circuit output file. The .FOUR statement produces outputs directly without the use of .PRINT, .PLOT, or .PROBE statements (see Figure 2.7).

The basic form of the .FOUR statement is

.FOUR <fundamental frequency> <output variable>

The parameter *<fundamental frequency>* is the fundamental frequency of the signal on which the Fourier decomposition is based. The *<output variable>* is as defined in Table 2.8 and can be a voltage or a current. The .FOUR statement can be used only with a .TRAN statement. PSpice uses the results of the transient analysis for the specified output variable(s). From these voltages and/or currents the Fourier components are calculated. The .FOUR statement produces only a tabular output, as shown in the following example.

Example 3.4.1

To demonstrate PSpice Fourier analysis capability, the lowpass filter considered in Example 3.3.2 will be used. Here, PSpice will be used to calculate the Fourier components of the filter input and output signals. The input signal is a square waveform of 2V amplitude and 125kHz frequency, as shown in Figure 3.13. To perform the Fourier analysis, the following statement

$$.FOUR\ 125kHz\ V(1),\ V(4)$$

must be included in Listing 3.2. This statement instructs PSpice to calculate the Fourier components of the filter input voltage signal, V(1), and the output voltage signal, V(4). The decomposition of the analysis is based on a 125kHz frequency (the input signal frequency). Running PSpice produces the Fourier components of each signal as shown in Box 3.1. These results are tabulated in the circuit output file. Box 3.1a shows the amplitude and phase values of the Fourier components of the input signal. This includes the DC, fundamental (or first), and second to ninth harmonics. The table shows the expected Fourier component values of a square waveform (i.e., odd harmonics only). PSpice outputs values relative to the fundamental component as well as the absolute values as shown in Box 3.1. Note that the output file also contains a calculation called "total harmonic distortion". Box 3.1b shows the Fourier components of the filter output signal. As expected, the harmonics have been attenuated according to the filter characteristics.

3.5 Fourier analysis (graphics mode)

It is also possible to obtain the Fourier components of a signal in graphic format using *Probe*. This is achieved by first displaying the required waveform in the time domain (transient analysis), select "X-axis" from the *Probe* menu and then "Fourier" from the submenu. It is important to specify the required output variable(s) in the .PROBE statement (see Table 2.8).

Example 3.5.1

To illustrate the use of *Probe* to generate the frequency spectrum of signals, consider again the filter used in Example 3.3.2. Having run Listing 3.2, the input and output signals, V(1) and V(4) respectively can now be plotted (Figure 3.13). Selecting "X-axis" from the main *Probe* menu, and then "Fourier" from the submenu, now allows the frequency spectrum of both signals to be obtained. The spectra of both signals are shown in Figure 3.16. As expected, they are in good agreement with the tabulated results given in Box 3.1.

Box 3.1a Part of PSpice Output File of Example 3.4.1

**** FOURIER ANALYSIS			TEMPERATURE = 27.000 DEG C	

****

FOURIER COMPONENTS OF TRANSIENT RESPONSE V(1)

DC COMPONENT =   2.506266E−03

HARMONIC NO	FREQUENCY (HZ)	FOURIER COMPONENT	NORMALIZED COMPONENT	PHASE (DEG)	NORMALIZED PHASE (DEG)
1	1.250E+05	1.273E+00	1.000E+00	−6.767E−01	0.000E+00
2	2.500E+05	5.013E−03	3.937E−03	8.865E+01	8.932E+01
3	3.750E+05	4.244E−01	3.333E−01	−2.030E+00	−1.353E+00
4	5.000E+05	5.013E−03	3.937E−03	8.729E+01	8.797E+01
5	6.250E+05	2.547E−01	2.000E−01	−3.383E+00	−2.707E+00
6	7.500E+05	5.014E−03	3.938E−03	8.594E+01	8.662E+01
7	8.750E+05	1.819E−01	1.429E−01	−4.737E+00	−4.060E+00
8	1.000E+06	5.015E−03	3.939E−03	8.459E+01	8.526E+01
9	1.125E+06	1.415E−01	1.111E−01	−6.090E+00	−5.414E+00

TOTAL HARMONIC DISTORTION =   4.288901E+01 PERCENT

Box 3.1b Part of PSpice Output File of Example 3.4.1

```
**** FOURIER ANALYSIS                    TEMPERATURE = 27.000 DEG C
******************************************************************************

FOURIER COMPONENTS OF TRANSIENT RESPONSE V(4)

DC COMPONENT =    1.254829E-03
```

HARMONIC NO	FREQUENCY (HZ)	FOURIER COMPONENT	NORMALIZED COMPONENT	PHASE (DEG)	NORMALIZED PHASE (DEG)
1	1.250E+05	6.364E-01	1.000E+00	-1.571E+01	0.000E+00
2	2.500E+05	2.512E-03	3.947E-03	5.862E+01	7.433E+01
3	3.750E+05	2.120E-01	3.331E-01	-4.814E+01	-3.243E+01
4	5.000E+05	2.509E-03	3.942E-03	2.468E+01	4.039E+01
5	6.250E+05	1.274E-01	2.001E-01	-8.454E+01	-6.884E+01
6	7.500E+05	2.515E-03	3.952E-03	-1.525E+01	4.612E-01
7	8.750E+05	9.089E-02	1.428E-01	-1.306E+02	-1.149E+02
8	1.000E+06	2.501E-03	3.929E-03	-7.218E+01	-5.648E+01
9	1.125E+06	6.359E-02	9.992E-02	1.488E+02	1.645E+02

```
TOTAL HARMONIC DISTORTION =    4.259478E+01 PERCENT
```

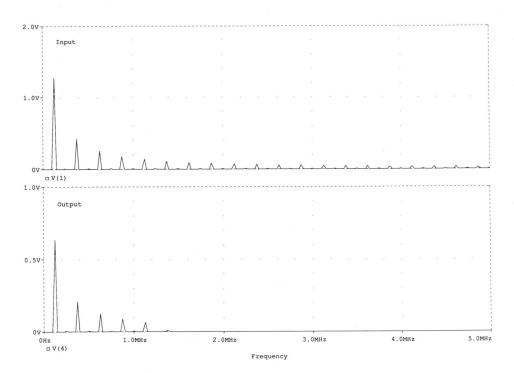

Figure 3.16 Top plot: Frequency spectrum of a 125kHz square waveform. Bottom plot: Frequency spectrum of a 125kHz square waveform having passed through a 1MHz filter (Figure 3.14).

It should be mentioned that the "ringing" noticed on the output signal (Figure 3.13) is due to a combination of the truncation of the Fourier series (Figure 3.16, bottom plot) and the filter nonlinear phase response (Figure 3.15). To summarise, Figure 3.17 shows the various options available to perform Fourier analysis.

The algorithm used in *Probe* to perform Fourier transform is based on the Fast Fourier Transform[4], or (FFT). This algorithm has two requirements that must be met: the time intervals between the individual calculations should be equally spaced, and the number of data points must be a power of 2. So, before the transform is done, *Probe* creates a new set of data points based on the points created by PSpice using an interpolation method. Figure 3.16 was produced using the following .TRAN statement:

.TRAN 0.03us 60us 0us 0.03us

which means the analysis has created 2000 data points or (0.03μs*2000 = 60μs). *Probe* will round this number up to 2048. *Probe* also creates a set

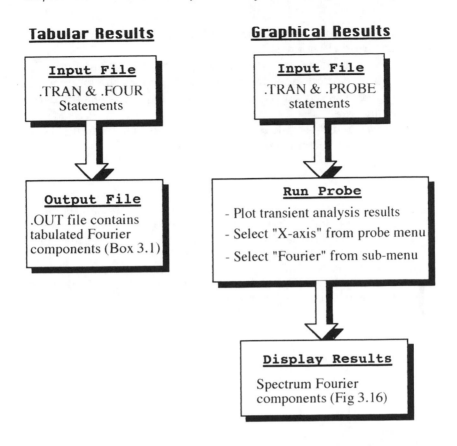

Figure 3.17   Options available to perform Fourier analysis.

of equally spaced data points, and in this case they are 60μs/2048, or
0.0293μs, apart. The resolution of the frequency spectrum shown in
Figure 3.16 is determined by the final time value on the .TRAN statement. In this case it is 60μs or 16.67kHz. To improve the resolution, the
final time value must be increased. This however increases the simulation time and the number of data points that might lead PSpice to report
the error message "Not enough memory for Fourier". It should be
noted that the interpolation method used to obtain equally spaced data
points is the reason why there is sometimes a slight difference between
the results obtained using the .FOUR statement and *Probe*.

3.6   Chapter summary

- PSpice has five time-dependent voltage and current sources.
  These are exponential (EXP), pulse (PULSE), sinusoidal (SIN),

single-frequency frequency-modulated (SFFM), and piece-wise linear (PWL) sources.
- Two statements are required to perform transient analysis. The first is a description statement, which defines the independent source. The second is a .TRAN statement, which sets the period of analysis.
- PSpice and *Probe* calculate and plot the Fourier components of a voltage or current signal (see Figure 3.17).

References

1. MicroSim Corporation, *Circuit Analysis — Reference Manual* (The Design Center), Version 5.3, Irvine, CA, January 1993, 134.
2. Bogart, F.B., *Electronics Devices and Circuits*, 2nd edition, Merrill Publishing Company, Columbus, OH, 1990, pp. 737-738.
3. Saal, R., *Handbook of Filter Design*, AEG Telefunken, 1979.
4. Rabiner, L.R., & Gold, R., *Theory and Applications of Digital Signal Processing*, Prentice Hall, Englewood Cliffs, NJ, 1985.

chapter four

Dependent sources (linear and nonlinear)

This chapter covers linear dependent sources that can be used to develop simple models of active elements such as amplifiers and transistors. The chapter also discusses nonlinear (polynomial) dependent sources. These sources can be used to model circuit operations such as "summers" and "multipliers" in block-diagram format. Worked examples of various linear and polynomial sources will be given.

4.1 Linear dependent sources

Active devices such as amplifiers can only be modelled by the use of "dependent sources". Unlike the independent sources discussed in Chapter 2, dependent sources have their output controlled by a number of inputs. PSpice has four dependent (controlled) sources: voltage-controlled voltage source (VCVS), voltage-controlled current source (VCCS), current-controlled current source (CCCS), and current-controlled voltage source (CCVS). Each of these dependent sources is recognised by PSpice as a component starting with the letter E, G, F, and H, respectively. Table 4.1 summarizes the PSpice linear dependent sources.

4.1.1 Linear voltage-controlled voltage source (VCVS)

The output voltage of an ideal op-amp (Figure 4.1) is given by

$$V_{out} = A_v(V_+ - V_-)$$

where V_{out} is the output voltage, A_v is the op-amp open loop gain, V_+ is the voltage at the noninverting input, and V_- is the voltage at the inverting input.

This equation shows that the op-amp output voltage is dependent on the voltage difference between its inputs multiplied by the open

Table 4.1 PSpice Linear Dependent Sources

Letter	Description
E	Voltage-controlled voltage source (VCVS)
G	Voltage-controlled current source (VCCS)
F	Current-controlled current source (CCCS)
H	Current-controlled voltage source (CCVS)

loop gain. An ideal op-amp (infinite input impedance, zero output impedance, and infinite gain vs. frequency) can therefore be modelled using a voltage-controlled voltage source (VCVS). The basic form of a VCVS statement is

> *E<name> <+ node> <– node>*
> *+ <+ controlling node> <- controlling node> <gain value>*

where *E* is the PSpice symbol for a VCVS, and *<name>* is the VCVS name, which can be up to eight characters long. The parameters *<+ node>* and *<– node>* are the output nodes of the VCVS. The parameters *<+ controlling node>* and *<– controlling node>* are the input nodes of the VCVS. The parameter *<gain value>* is the gain of the VCVS. To illustrate the use of the VCVS in describing ideal op-amps, consider the following example.

Example 4.1
Use PSpice to obtain the frequency response of the inverting amplifier circuit shown in Figure 4.2 over the frequency range of 1Hz to 10MHz.
    The PSpice input file of the amplifier is shown in Listing 4.1a. The op-amp has been called "Eop-amp" with the inverting input at node 2

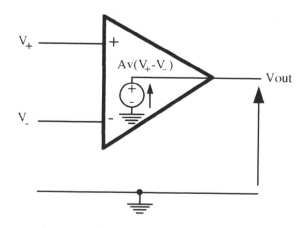

Figure 4.1 Equivalent circuit of an ideal op-amp.

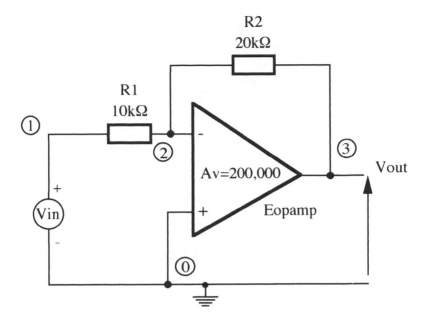

Figure 4.2   Circuit of Example 4.1.

and the noninverting input at node 0. The VCVS statement requires positive and negative output nodes to be specified. Most op-amps are single ended, and in this example the positive output of the amplifier is at node 3 and the negative output is at node 0, or ground. It has been assumed that the op-amp has a gain of 200,000. The simulated frequency response of the ideal amplifier circuit is shown as part of Figure 4.3. This shows the amplifier has a gain of 6dB, phase shift of 180°, and infinite bandwidth, as expected.

Listing 4.1a   PSpice Input File of Example 4.1

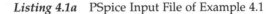

```
Ideal Inverting Amplifier (Figure 4.2)
Vin 1 0 AC 1
.AC LIN 700 1 10E6
**
R1 1 2 10K
R2 2 3 20K
**
Eop-amp 3 0 0 2 200E3    ; ideal op-amp description
**
.PROBE V(3)
.END
```

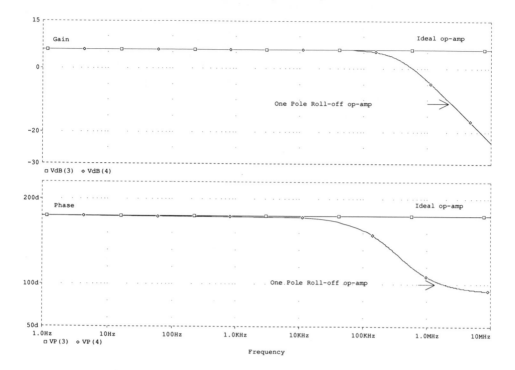

Figure 4.3 Ideal and practical response of an inverting amplifier (gain = 2).

A simple, practical model to show a typical op-amp gain roll-off characteristic can be developed by including an RC network within the VCVS, as shown in Figure 4.4. This op-amp model is called the single-pole roll-off model[1], has a slope of –20dB per decade and a phase shift of –90° at high frequency. The RC values are related to the amplifier bandwidth by

$$\text{Amplifier bandwidth} = 1/(2\pi RC)$$

The PSpice input file of the circuit in Figure 4.4 is given in Listing 4.1b. Figure 4.3 shows the modelled amplifier frequency response, assuming that the op-amp has a gain-bandwidth product of 1MHz and an open loop gain of 200,000. Figure 4.3 was obtained as a result of combining Listings 4.1a and 4.1b as one file and running a single analysis. It should be noted that this is a simple op-amp model and that a comprehensive model is discussed in Chapter 6.

4.1.2 Linear voltage-controlled current source (VCCS)

The relationship between the input and the output of an ideal operational transconductance amplifier (OTA), Figure 4.5, is given by

$$I_{out} = g_m (V_+ - V_-)$$

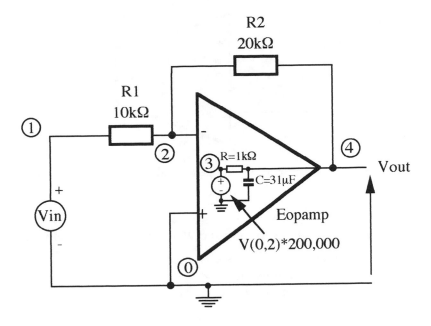

Figure 4.4 Practical amplifier circuit.

where I_{out} is the output current, g_m is the transconductance of the amplifier in siemens (S), V_+ is the voltage at the noninverting input, and V_- is the voltage at the inverting input. It has been assumed that the OTA has infinite input and output impedance.

This equation shows that the OTA output current is controlled by the input voltage multiplied by its transconductance value. Therefore,

Listing 4.1b PSpice Input File of Circuit (Figure 4.4)

```
Practical Inverting Amplifier (Figure 4.4)
*
Vin 1 0 AC 1
.AC LIN 700 1 10E6
**
R1 1 2 10K
R2 2 4 20K
*
R 3 4 1K
C 4 0 31uF; op-amp pole @ 5Hz
*
Eopamp 3 0 0 2 200E3; ideal op-amp description
*
.PROBE V(4)
*
.END
```

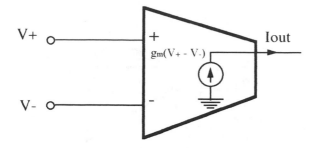

Figure 4.5 Equivalent circuit of an operational transconductance amplifier (OTA).

an ideal OTA can be described by a linear voltage-controlled current source (VCCS). The basic form of the VCCS description statement is

> G<name> <+ node> <- node>
> + <+ controlling node> <- controlling node>
> + <transconductance value>

where G is the PSpice symbol for a VCCS and <name> is the VCCS name, which can be up to eight characters long. The <+ node> and <- node> are the output nodes, and the <+ controlling node> and <- controlling node> are the input nodes of the VCCS. The <transconductance value> is that of the VCCS.

    To illustrate the use of the VCCS in describing ideal OTAs, consider the following example.

Example 4.2
OTAs are commercially available as ICs from a number of manufacturers[2-3], and they often find applications in amplifier, filter, and oscillator circuit design[4]. Figure 4.6 shows a first-order lowpass filter where the voltage transfer function (output/input) is given by

$$\frac{V_{out}}{V_{input}} = \frac{g_m}{sC + g_m}$$

and the –3dB cutoff frequency of this filter is

$$F_{-3dB} = \frac{g_m}{2\pi C}$$

    Assuming that $F_{-3dB} = 1.59$kHz and $C = 0.1\mu$F, this gives $g_m = 1$E-3 (siemens). Note that OTA-based filters use g_ms in the place of resistors in conventional op-amp-based filters.

    The PSpice input file of the OTA lowpass filter is given in Listing 4.2. The resistor, RL, is included in the file so that no floating nodes are present. The high value of this resistor is chosen so that it does not affect

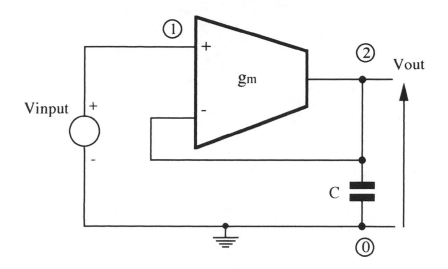

Figure 4.6 Circuit of Example 4.2.

the filter operation. The simulated frequency response of the OTA filter is shown in Figure 4.7.

4.1.3 Linear current-controlled current source (CCCS)

The relationship between the base current (I_B) and the collector current (I_C) of a bipolar transistor is given by

$$\beta = I_C/I_B$$

This equation shows that the I_C of a bipolar transistor is controlled by the base current, I_B. A bipolar transistor is therefore a CCCS. The basic form of the CCCS description statement is

$$F<name> <+\ node> <-\ node> <Vxxxx> <current\ gain>$$

Listing 4.2 PSpice Input File of Example 4.2

```
1st-order OTA lowpass filter (Figure 4.6)
Vin 1 0 AC 1
.AC LIN 500 1 50K
**
C 2 0 0.1uF
RL 2 0 1E9              ; see text
G1 2 0 1 2 1E-3         ; OTA ideal model
**
.PROBE V(2)
.END
```

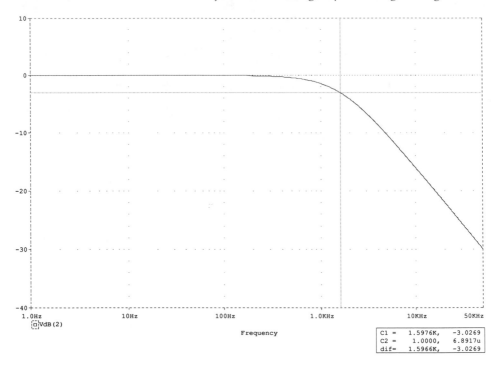

Figure 4.7   Frequency response of Example 4.2.

where *F* is the PSpice symbol for a CCCS and *<name>* is the CCCS name,
which can be up to eight characters long. The *<+ node>* and *<– node>* are
the output nodes of the CCCS. The parameter *<Vxxxx>* is the name of
an independent DC voltage source through which the current controlling the CCCS flows. This voltage source (with a value of 0V) must be
inserted in the circuit where the controlling current flows. This voltage
source has no effect on the circuit operation. It should be noted that this
is a special case where PSpice measures current by the use of a DC
voltage source with a value of 0V. Measuring current through components is usually achieved as discussed in Table 2.8.

The parameter *<current gain>* defines the gain of the CCCS. To
illustrate the use of the CCCS, consider the following example.

Example 4.3
Describe the operation of the transistor amplifier circuit (Figure 4.8a)
using CCCS. The equivalent amplifier circuit using a CCCS is shown in
Figure 4.8b.

The controlling current (base current) is measured using a DC
voltage source, Vib, connected between nodes 2 and 3 and should be set
to 0V as shown in the following statement:

Vib 2 3 DC 0V

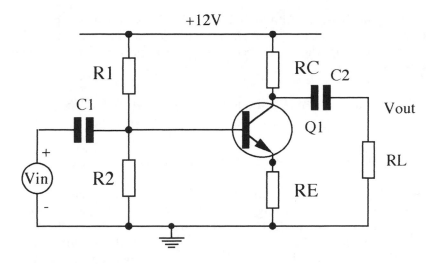

Figure 4.8a   Common emitter amplifier circuit.

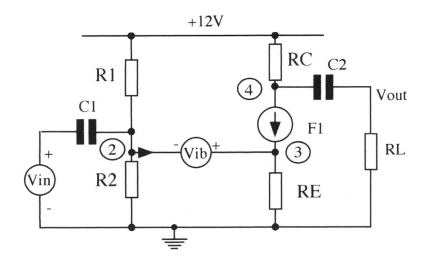

Figure 4.8b   Circuit of Figure 4.8a with the transistor being described using CCCS.

Output current is described by the following CCCS statement:

F1 4 3 Vib 100

This statement shows that the current in the CCCS (F1) connected between nodes 4 and 3 has a magnitude of 100 times the current flowing through the voltage source, Vib. It has been assumed that the transistor current gain, b, is 100. Although describing bipolar transistor operation using CCCS is more complex than using the Q description statement for discrete designs, the CCCS becomes more important for functional simulation designs.

4.1.4   Linear current-controlled voltage source (CCVS)

The basic form of the CCVS is

H<name> <+ node> <– node> <Vxxxx> <transresistance value>

where *H* is the PSpice symbol for a CCVS and *<name>* is the CCVS name, which can be up to eight characters long. The *<+ node> <– node>* are the output nodes of the CCVS. The parameter *<Vxxxx>* is the name of an independent DC voltage source through which the current controls the CCVS flow. This voltage source, which must be inserted in the circuit where the control current flows, has a value of 0V so that it has no effect on the circuit operation. The parameter *<transresistance value>* defines the transresistance of the CCVS in ohms.

To demonstrate a CCVS, it is necessary to contrive such a function as shown in Figure 4.9a. The output voltage of this amplifier circuit is

$$V_{out} = RF * I1$$

This shows that the amplifier circuit is a current-to-voltage converter. Therefore, the operation of this circuit can be described using a CCVS. The equivalent amplifier circuit using a CCVS is shown in Figure 4.9b.

The controlling current, I1, is measured using a DC independent voltage source, V1, connected between nodes 1 and 2 and set to 0V as shown in the following statement:

V1 1 2 DC 0V

Note that this voltage source is introduced only for measurement purposes.

The CCVS is described using the following statement:

H 3 2 V1 1K

where 1K represents the value of the feedback resistor. It should be noted that this amplifier circuit (Figure 4.9a) would be better described using a VCVS. Both CCCS and CCVS often find applications in functional simulation designs.

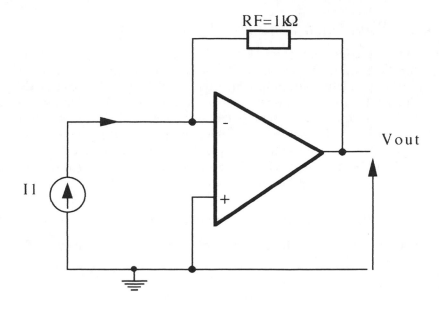

Figure 4.9a   Current-to-voltage converter.

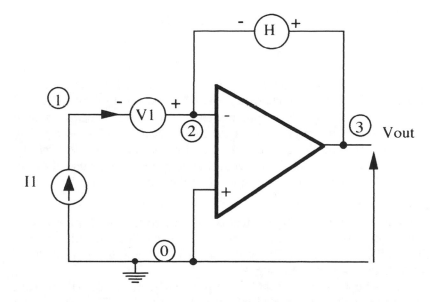

Figure 4.9b   Circuit of Figure 4.9a described using CCVS.

4.2 Polynomial nonlinear dependent sources

So far, the controlled sources have been linearly dependent on the controlling inputs. PSpice allows the user to specify polynomial (nonlinear) dependent sources in which the controlling inputs can be described in the form of a polynomial. Such sources are useful since they can be used to describe circuit operations such as summers, and multipliers in block-diagram format. PSpice supports four polynomial dependent sources, similar to the dependent sources. These are VCVS, VCCS, CCCS, and CCVS. Only the polynomial VCVS will be discussed here, since the applications of the other sources are similar.

4.2.1 One-dimensional polynomial VCVS

The output of the one-dimensional polynomial VCVS is given by the equation

$$V_{out} = P_0 + P_1 V_1 + P_2 V_1^2 + P_3 V_1^3 + \ldots$$

where V_{out} is the output voltage of the VCVS, P_0, P_1, P_2,..., are the polynomial coefficients, and V_1 is the controlling input voltage. The basic form of the one-dimensional polynomial VCVS is

> E<name> <+ node> <– node> POLY(1)
> + (<+ controlling node>,<- controlling node>)
> + (<polynomial coefficient values>)

where *E* is the PSpice symbol for a VCVS and *<name>* is the VCVS name, which can be up to eight characters long. The parameters *<+ node>* and *<– node>* are the VCVS output nodes, and *POLY(1)* specifies a one-dimensional polynomial. The *<+ controlling node>* and *<– controlling node>* are the controlling input nodes, and *<polynomial coefficient values>* represents the polynomial coefficients. To demonstrate the use of a one-dimensional VCVS source, consider the following example.

Example 4.4
Simulate the operation of an ideal two-input voltage multiplier circuit when both inputs are 1V, 1kHz sine waveforms. The block diagram of this multiplier circuit is shown in Figure 4.10. The PSpice input file of this circuit is given in Listing 4.3.

It has been assumed that both inputs of the multiplier (Emult) are connected to node 1 and that the output is at node 2. The polynomial coefficients are set according to the one-dimensional polynomial equation given above, where the first and second coefficients are set to

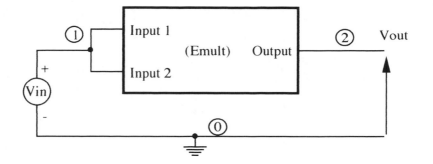

Figure 4.10 Circuit of Example 4.4.

"zero" and the third coefficient is set to "one", as shown in the listing. The simulated input and output waveform of the multiplier circuit is shown in Figure 4.11.

4.2.2 Two-dimensional polynomial VCVS

The output of two-dimensional VCVS, V_{out}, is given by the equation

$$V_{out} = P_0 + P_1V_1 + P_2V_2 + P_3V_1^2 + P_4V_1V_2 + P_5V_2^2 + \ldots$$

where $P_0, P_1, \ldots$, are the polynomial coefficients, and V_1 and V_2 are the two controlling inputs. The basic form of the two-dimensional VCVS source is

> E<name> <+ node> <– node> POLY (2)
> + (<+ controlling input1>,<- controlling input1>)
> + (<+ controlling input2>,<- controlling input2>)
>   (<polynomial coefficient values>)

Listing 4.3 PSpice Input File of Example 4.4

```
Ideal Multiplier Circuit  (Figure 4.10)
*
Vin 1 0 SIN (0V 1V 1kHz)             ; 1V, 1kHz sine wave
.TRAN 0.01ms 3ms 0ms 0.01ms
**
Emult 2 0 POLY(1)  (1,0)  (0,0,1)    ; one-dimensional VCVS
**
.PROBE V(1), V(2)
.END
```

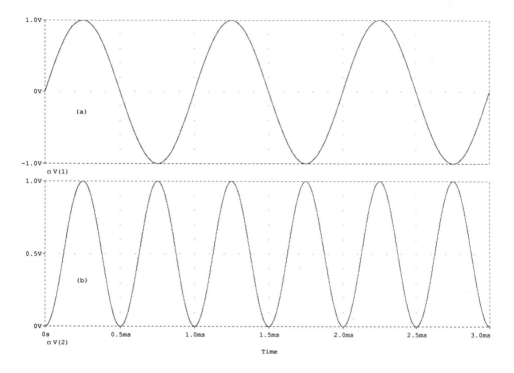

Figure 4.11   Transient response of Example 4.4: (a) input; (b) output.

which is similar to the one-dimensional VCVS, expect that there are two sets of controlling input nodes and the number after the parameter POLY has been changed to 2. To illustrate the use of a two-dimensional VCVS, consider the following example.

Example 4.5
Simulate the operation of an ideal two-input voltage multiplier circuit when one input is a 1V, 10kHz sine wave, and the second input is a 1V, 300kHz sine wave. The block diagram of this multiplier circuit is shown in Figure 4.12.

The PSpice input file of this circuit is given in Listing 4.4. The polynomial coefficients are set according to the two-dimensional VCVS equation, where the first, second, third, and fourth coefficients are set to "zero", and the fifth coefficient is set to "one". The simulated inputs and the output signals are shown in Figure 4.13.

Higher-dimension polynomial VCVS sources are possible, but they are usually limited in applications and difficult to implement. An easier and more efficient way of implementing polynomial controlled sources can be achieved using the VALUE option of the analogue behavioural modelling, shown later in Chapter 8 (Section 8.3).

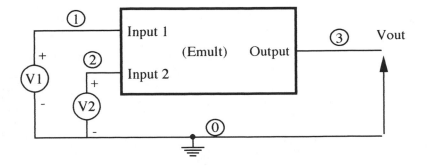

Figure 4.12 Circuit of Example 4.5.

4.3 Effective use of dependent sources

A specific speech-coding application requires a lowpass filter with the following specifications:

> Passband edge=3.6kHz
> Stopband edge=4kHz
> Stopband attenuation≥60dB

One method usually used to design such a sharp filter is the frequency dependent negative resistor[5] (FDNR). This method basically involves changing an LC passive filter into its active counterpart. Figure 4.14 shows the required filter[6], and it is a ninth-order elliptic circuit.

Having designed the filter, PSpice can now be used to help the designer

1. Check to see if the filter functions as designed by using ideal components
2. Predict how the filter performs when built by using modelled (practical) components

Listing 4.4 PSpice Input File of Example 4.5

```
Ideal Multiplier Circuit (Figure 4.12)
*
V1 1 0 SIN (0V 1V 10kHz)      ; input signal 1
V2 2 0 SIN (0V 1V 300kHz)     ; input signal 2
.TRAN 0.1us 200us 0us 0.1us   ; transient analysis range
*
* Two dimensional VCVS
Emult 3 0 POLY(2) (1,0) (2,0) (0 0 0 0 1)
*
.PROBE V(1), V(2), V(3)        ; graphic output
*
.END
```

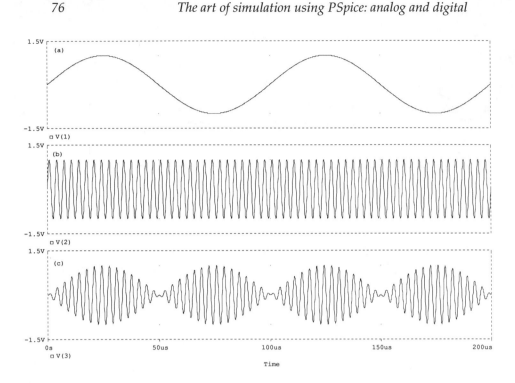

Figure 4.13   Transient response of Example 4.5: (a) signal 1; (b) signal 2; (c) output.

The two simulations above require different levels of component modelling; for simplicity, the filter resistors and capacitors will be assumed perfect (component tolerances are considered in Chapter 7). The first simulation tests whether the filter components were calculated correctly. At this stage, ideal op-amp models based on VCVS are sufficient. The PSpice input file of the FDNR filter is given in Listing 4.5. The simulated frequency response of the filter is shown in Figure 4.15, which confirms the filter functions as designed. The second simulation involves using more complex (practical) op-amp models to predict the

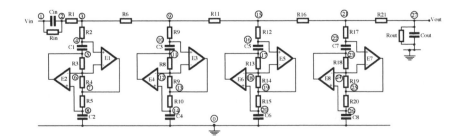

Figure 4.14   The 3.6kHz lowpass active filter.

Listing 4.5   PSpice Input File of Circuit (Figure 4.14)

```
3.6kHz lowpass FDNR filter (Figure 4.14)
*
VIN 1 0 AC 1
.AC LIN 800 1 10kHz
*
* Resistors
*
Rin 1 2 1meg
R1 2 3 6.085k
R2 3 4 0.649k
R3 5 6 5.38k
R4 6 7 5.38k
R5 7 8 5.38k
R6 3 9 7.555k
R7 9 10 4.082k
R8 11 12 3.39k
R9 12 13 3.39k
R10 13 14 3.39k
R11 9 15 5.267k
R12 15 16 5.602k
R13 17 18 2.78k
R14 18 19 2.78k
R15 19 20 2.78k
R16 15 21 6.022k
R17 21 22 2.45k
R18 23 24 3.95k
R19 24 25 3.95k
R20 25 26 3.95k
R21 21 27 4.755k
Rout 27 0 1meg
*
* capacitors
Cin 1 2 10nF
C1 4 5 10nF
C2 8 0 10nF
C3 10 11 10nF
C4 14 0 10nF
C5 16 17 10nF
C6 20 0 10nF
C7 22 23 10nF
C8 26 0 10nF
Cout 27 0 10nF
*
* ideal op-amps, op-amp gain = 200,000
E1 7 0 4 6 200k
E2 5 0 8 6 200k
E3 13 0 10 12 200k
E4 11 0 14 12 200k
E5 19 0 16 18 200k
E6 17 0 20 18 200k
E7 25 0 22 24 200k
E8 23 0 26 24 200k
*
.PROBE V(27)
.END
```

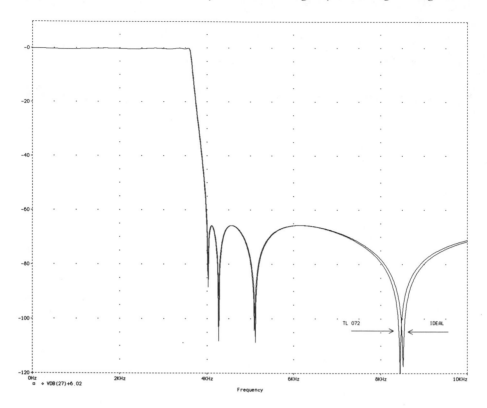

Figure 4.15 Frequency response of circuit (Figure 4.14) using ideal and TL072 op-amps.

filter response when it is built. Semiconductor manufacturers usually provide op-amp models, and the accuracy and complexity of such models often vary depending on the manufacturer. Usually, the more complex the model is, the longer the simulation takes, and there is a greater chance that PSpice will fail to produce a result due to convergence problems. Assume that the filter will be realised using TL072 op-amps. A model for this op-amp can be found in the Texas Instruments library supplied with PSpice (see Chapter 6, Section 6.5). Figure 4.15 shows the frequency response of the filter using the TL072 op-amp model. As can be seen, there is a very small difference between the ideal and the modelled filter frequency response. This result shows the importance of not overspecifying the op-amp models. It should be noted that the simulation time of the ideal filter was 18 s, while that of the modelled filter was 83 s. Both simulations were performed on a 486DX, 33MHz IBM PC.

This example has shown how PSpice can help the designer during the various stages of circuit design and implementation, and it highlights the importance of using appropriate component models. Component modelling is dealt with in Chapter 6.

4.4   Chapter summary

- PSpice has four linear dependent sources. These are VCVS, VCCS, CCCS, and CCVS. Such sources can be used to describe the operation of simple active elements such as amplifiers.
- PSpice supports four polynomial (nonlinear) dependent sources. These sources can be used to describe active circuit operations such as multipliers and summers in block-diagram format.
- The designer has to decide on the complexity of active component models so that the required simulation result is achieved, in terms of simulation time and accuracy of results.

References

1. Van Valkenburg, M.E., *Analogue Filter Design*, Holt, Rinehart & Winston, New York, 1982.
2. National Semiconductor Corporation, *LM13600 (Dual OTA with Linearsing Diode and Buffer)*, Linear Devices Data Book.
3. Burr-Brown Corporation, OPA660 *(Wide Bandwidth OTA and Buffer)*, Integrated Circuits Data Book Supplement, Vol. 33C, 1992.
4. Geiger, R.L., & Sanchez-Sinencio, E., "Active Filter Design Using Operational Transconductance Amplifiers: A Tutorial", *IEEE Circuits and Devices Magazine*, Vol. 1, 1985, pp. 20-32.
5. Bruton, L.T., "Network Transfer Functions Using the Concept of Frequency Dependent Negative Resistance", *Trans Circuit Theory*, Vol. CT-16, 1969, pp. 406-408.
6. Al-Hashimi, B.M., "Building Bricks into Brick Wall Filters", *Electronics World + Wireless World*, June 1992, pp. 461-464.

chapter five

Subcircuits and efficient use of PSpice

Electronic circuit designs often involve the use of similar, or even identical, blocks of circuitry. PSpice provides the user with a powerful set of commands to allow repetitive circuitry or design iteration to be done efficiently. It is possible to "program" PSpice, and there are three main statements available to do this. This chapter discusses such PSpice statements and demonstrates their use through worked examples.

5.1 The .PARAM statement

Sometimes it is more convenient to set component values using mathematical expressions in place of numerical values. PSpice allows the user to do this with the .PARAM statement, and the basic form of this statement is

$$.PARAM <<name> = <value>>$$
$$.PARAM <<name> = {<expression>}>$$

where parameter name (*<name>*) can be any arbitrary set of characters allowed by PSpice. The parameter value (*<value>*) may be a constant or mathematical expression. The mathematical expression can have any of the functions shown in Table 5.1, and the four standard operators (+, −, *,/) may be used. If a mathematical expression is used, the <value> statement must be enclosed in brackets of the type { }. For example,

$$.PARAM R1 = 1k$$
$$.PARAM R2 = {2*R1}$$

The first statement sets the value R1 to 1kΩ, and the second statement sets R2 to 2*R1 = 2kΩ.

Table 5.1   PSpice Arithmetic Functions

Function	Meaning	Comment
ABS(x)	$\lvert x \rvert$	Absolute value of x
SQRT(x)	$x^{1/2}$	Square root of x
EXP(x)	e^x	e to the x power
LOG(x)	ln (x)	Log base e of x
LOG10(x)	log (x)	Log base 10 of x
PWR(x,y)	$\lvert x \rvert^y$	Absolute value of x to the y power
SIN(x)	sin (x)	(x in radians)
COS(x)	cos (x)	(x in radians)
TAN(x)	tan (x)	(x in radians)
ARCTAN(x)	$\tan^{-1}$ (x)	Result in radians

5.2   Subcircuits

In order to describe repetitive blocks of circuitry, the subcircuit approach offers the designer flexibility and ease of use. In this approach, by writing a single PSpice statement, a repetitive or an often-used circuit block can be defined once and then used each time the circuit block is required. The concept of subcircuits is similar to that of subroutines in conventional computer programming.

To define a subcircuit, the .SUBCKT statement is required. The basic form of this statement is

.SUBCKT <name> <nodes> [PARAMS: <<name> = <value>>]

where *<name>* represents any name chosen to identify the subcircuit, and *<nodes>* represents the external nodes of the subcircuit. The nodes internal to a subcircuit are local to that subcircuit, which means that it is possible to duplicate or use node numbers in a subcircuit that has already been used in the main circuit input file.

To increase the flexibility of the subcircuit, the user is allowed to pass a number of optional parameters into the subcircuit. The names and values of these parameters are set by *<name>* and *<value>*.

The .SUBCKT statement is the first statement in a subcircuit definition, followed by description statements that define the components in the subcircuit. The last statement must be an .ENDS statement, which marks the end of a subcircuit definition. The basic form of this statement is

.ENDS [subcircuit name]

It is good practice to repeat the subcircuit name although this is not strictly necessary. So, the format of a subcircuit is as follows:

.SUBCKT statement
Circuit description statements
.ENDS statement

5.2.1   Calling subcircuits

When a subcircuit is called, it is essential that a unique name is used for each separate call; this name is then recognised by PSpice as a device. The statement required in the main circuit input file is as follows:

X<name> [nodes] <subcircuit name> [PARAMS: <<name> = <value>>]

where *X<name>* is the device name of the subcircuit, which can be up to eight characters long. This statement causes the referenced subcircuit to be inserted into the main circuit with the X statement nodes matching the nodes used in the .SUBCKT statement. There must be the same number of nodes and parameters in the X statement as in the .SUBCKT statement, and it is essential that they be in the same order. The *<subcircuit name>* is the name of the subcircuit to be inserted as previously defined. This means that once a subcircuit has been defined, it can be called as a device having a name that starts with an "X".

Example 5.1
To illustrate the development of subcircuits and their use, consider the lowpass FDNR active filter shown in Figure 5.1. This circuit was described in Chapter 4, using the basic description statements as shown in Listing 4.5, and is repeated here for convenience. Listing 5.1a shows that more than 40 description statements were used to define the filter components. Figure 5.1 shows that the filter circuit has four identical FDNR elements. To avoid entering the same component statements describing each FDNR element four times, the FDNR element should be defined as a subcircuit and then called as required.

Using a text editor, the FDNR element (Figure 5.2a) is defined as a subcircuit:

```
.SUBCKT  FDNR  1  7  PARAMS:  RA = 1,  R = 1,  C = 1
RA  1  2  {RA}
CA  2  3  {C}
RB  3  4  {R}
RC  4  5  {R}
RD  5  6  {R}
CB  6  7  {C}
E1  5  0  2  4  200K
E2  3  0  6  4  200K          ;  ideal  op-amp  description
.ENDS  FDNR
```

Listing 5.1a PSpice Input File of Example 5.1

```
3.6kHz lowpass FDNR filter (Figure 5.1)
VIN 1 0 AC 1
.AC LIN 800 1 10kHz
*
* Resistors
Rin 1 2 1meg
R1  2  3  6.085k
R2  3  4  0.649k
R3  5  6  5.38k
R4  6  7  5.38k
R5  7  8  5.38k
R6  3  9  7.555k
R7  9  10  4.082k
R8  11  12  3.39k
R9  12  13  3.39k
R10  13  14  3.39k
R11  9  15  5.267k
R12  15  16  5.602k
R13  17  18  2.78k
R14  18  19  2.78k
R15  19  20  2.78k
R16  15  21  6.022k
R17  21  22  2.45k
R18  23  24  3.95k
R19  24  25  3.95k
R20  25  26  3.95k
R21  21  27  4.755k
Rout 27 0 1meg
*
* capacitors
Cin 1 2 10nF
C1  4  5  10nF
C2  8  0  10nF
C3  10  11  10nF
C4  14  0  10nF
C5  16  17  10nF
C6  20  0  10nF
C7  22  23  10nF
C8  26  0  10nF
Cout 27 0 10nF
*
* ideal op-amps, op-amp gain = 200,000
E1  7  0  4  6  200k
E2  5  0  8  6  200k
E3  13  0  10  12  200k
E4  11  0  14  12  200k
E5  19  0  16  18  200k
E6  17  0  20  18  200k
E7  25  0  22  24  200k
E8  23  0  26  24  200k
*
.PROBE V(27)
.END
```

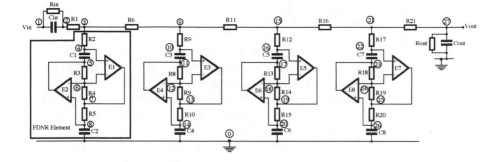

Figure 5.1   The 3.6kHz lowpass FDNR filter.

The subcircuit has been called FDNR, with its external nodes at 1 and 7 (Figure 5.2b). Note that even though it is obvious that node 7 will become 0V, that is, node 0, it is forbidden to use node 0 as an external node in the .SUBCKT statement. The FDNR subcircuit has three parameters, RA, R, and C, as shown in Figure 5.2a. The parameters RA, R, and C must be given initial values to satisfy the .SUBCKT statement

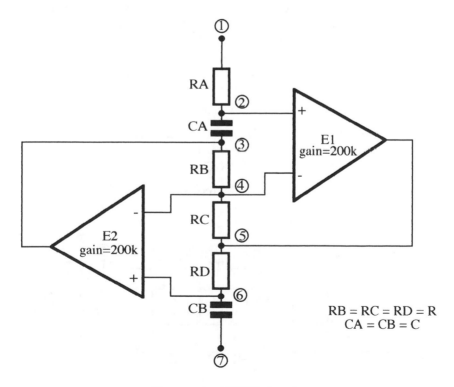

$$RB = RC = RD = R$$
$$CA = CB = C$$

Figure 5.2a   FDNR element.

Figure 5.2b   FDNR subcircuit.

requirement. In this case, these initial values are arbitrarily set to 1, since they will be changed to the required values when the subcircuit is called. For simplicity, it has been assumed that the op-amps are ideal (i.e., described using the E component) and have a gain of 200K.

Figure 5.3 shows the FDNR lowpass filter in a subcircuit notation. The PSpice input file of this circuit is given in Listing 5.1b. In this example, the first call statement (X1) connects subcircuit (FDNR) nodes 1 and 7 to main circuit nodes 3 and 0. The parameters RA, R, and C have been given the values of 0.649kΩ, 5.38kΩ, and 10nF, respectively (overriding the default values of 1). Similarly, X2 connects subcircuit nodes 1 and 7 to main circuit nodes 4 and 0, where RA, R, and C are equal to 4.082kΩ, 3.39kΩ and 10nF, respectively. Note that some of the subcircuit and the main circuit node numbers are the same. This is acceptable because the subcircuit nodes are local and PSpice does not confuse them with the global nodes in the main circuit. It is good practice to collect subcircuits in a user-defined file, which is refereed to using the PSpice command

.LIB <file name>

For example, in Listing 5.1b, the FDNR subcircuit is included in a file called <circuits>, which is referred to and has been assumed to be located on drive A using the statement

.LIB A:\circuits

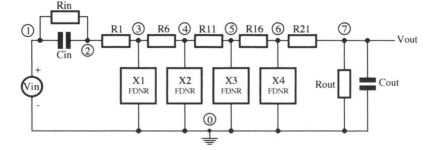

Figure 5.3   Lowpass filter (Figure 5.1) in FDNR subcircuit notation.

Listing 5.1b PSpice Input File of Example 5.1

```
3.6kHz FDNR lowpass filter (Figure 5.3)
**
.LIB A:\circuits          ; "FDNR" subcircuit location
VIN 1 0 AC 1
.AC LIN 600 1 10K
**
Cin 1 2 10nF
Cout 7 0 10nF
Rin 1 2 1MEG
Rout 7 0 1MEG
**
R1 2 3 6.085K
X1 3 0 FDNR PARAMS: RA = 0.649K, R = 5.38K, C = 10nF
**
R6 3 4 7.555K
X2 4 0 FDNR PARAMS: RA = 4.082K, R = 3.39K, C = 10nF
**
R11 4 5 5.267K
X3 5 0 FDNR PARAMS: RA = 5.602K, R = 2.78K, C = 10nF
**
R16 5 6 6.022K
X4 6 0 FDNR PARAMS: RA = 2.45K, R = 3.95K, C = 10nF
**
R21 6 7 4.755K
**
.PROBE V(7)
.END
```

5.2.2 Components scaling

Using mathematical expressions in the .PARAM statement allows the designer to scale component values. For example, once a filter has been designed, it often needs scaling in frequency. Scaling FDNR filters[1] is achieved by multiplying all the resistors (including the FDNR elements) by the factor (F1/F2), where F1 is the existing cutoff frequency and F2 is the new cutoff frequency, while keeping existing capacitor values unchanged. Thus if the 3.6kHz filter (Figure 5.1) is to be scaled to 7kHz, for example, the designer needs to perform the above simple scaling procedure and then enter the new set of component values. It is possible, however, to use PSpice to perform the scaling procedure as follows:

1. Set F1 and F2 using .PARAM statements.
2. Define the scaling factor (SF) = (F1/F2) as an expression using a .PARAM statement.
3. Multiply all filter resistance values by the scaling factor.

Listing 5.2   PSpice Input File of Example 5.2

```
7kHz FDNR lowpass filter
**
.LIB A:\circuits; "FDNR" subcircuit location
.OPTIONS LIST; provides components listing in .OUT file
**
VIN 1 0 AC 1
.AC LIN 600 1 20K
**
*
.PARAM F1 = 3.6kHz; existing cutoff frequency
.PARAM F2 = 7kHz; new cutoff frequency
.PARAM SF = {F1/F2}; scaling factor
**
*
Cin 1 2 10nF
Cout 7 0 10nF
Rin 1 2 1MEG
Rout 7 0 1MEG
**
R1 2 3 {6.085k*SF}
X1 3 0 FDNR PARAMS: RA = {0.649K*SF}, R = {5.38K*SF}, C = 10nF
**
R6 3 4 {7.555k*SF}
X2 4 0 FDNR PARAMS: RA = {4.082K*SF}, R = {3.39K*SF}, C = 10nF
**
R11 4 5 {5.267k*SF}
X3 5 0 FDNR PARAMS: RA = {5.602K*SF}, R = {2.78K*SF}, C = 10nF
**
R16 5 6 {6.022k*SF}
X4 6 0 FDNR PARAMS: RA = {2.45K*SF}, R = {3.95K*SF}, C = 10nF
**
R21 6 7 {4.755k*SF}
**
.PROBE V(7)
**
.END
```

To demonstrate the scaling procedure, consider the following example.

Example 5.2
Assume that we would like to scale the original 3.6kHz filter (Figure 5.1) to 7kHz. The modified PSpice input file of the FDNR filter is given in Listing 5.2, which includes the scaling procedure. It is essential to enclose the resistor values of the filter within brackets of the type { }. This is because the values of the resistors are described by expressions. The simulated frequency response of the scaled 7kHz filter is shown in Figure 5.4.

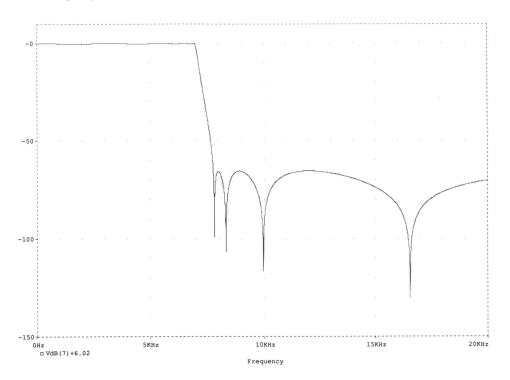

Figure 5.4   Frequency response of Example 5.2.

A components list of the 7kHz filter can be obtained by specifying the statement

.OPTIONS LIST

in the circuit input file. This option lists a summary of circuit elements. PSpice has other options that allow the user to print and suppress circuit and simulation information as required. These options are discussed in Appendix F. The 7kHz filter components are included in the circuit output file as shown in Box 5.1. Note that using this scaling procedure changes only the resistor values of the filter. The resistors and capacitors of each FDNR element are prefixed with the subcircuit device name. For example, "RA" of X1 becomes "X1.RA".

5.3   The .STEP statement

This statement allows the user to step the value of a signal source, a circuit component, or a temperature while running analyses for each step. The stepping can be linear or logarithmic or just a list of values. Stepping temperature values is demonstrated in Chapter 7 (Section 7.6).

Box 5.1   Part of PSpice Output File of Example 5.2.

```
*** CIRCUIT ELEMENT SUMMARY
****************************************************************

***  RESISTORS

NAME              VALUE
Rin               1.00E+06
Rout              1.00E+06
R1                3.13E+03
R6                3.89E+03
R11               2.71E+03
R16               3.10E+03
R21               2.45E+03
X1.RA             3.34E+02
X1.RB             2.77E+03
X1.RC             2.77E+03
X1.RD             2.77E+03
X2.RA             2.10E+03
X2.RB             1.74E+03
X2.RC             1.74E+03
X2.RD             1.74E+03
X3.RA             2.88E+03
X3.RB             1.43E+03
X3.RC             1.43E+03
X3.RD             1.43E+03
X4.RA             1.26E+03
X4.RB             2.03E+03
X4.RC             2.03E+03
X4.RD             2.03E+03

****  CAPACITORS

NAME              VALUE
Cin               1.00E-08
Cout              1.00E-08
X1.CA             1.00E-08
X1.CB             1.00E-08
X2.CA             1.00E-08
X2.CB             1.00E-08
X3.CA             1.00E-08
X3.CB             1.00E-08
X4.CA             1.00E-08
X4.CB             1.00E-08
```

Here we focus on the stepping of component values. The basic form of the .STEP statement is

.STEP PARAM <sweep variable name>
+ <start value> <end value> <increment value>

where the <sweep variable name> must be defined using a .PARAM statement. The .STEP command is similar in application to performing

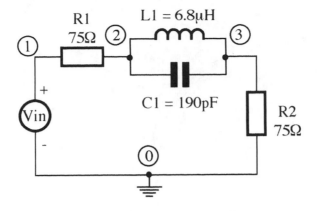

Figure 5.5   Circuit of Example 5.3.

Listing 5.3   PSpice Input File of Example 5.3

```
4.43MHz Notch Filter (Figure 5.5)
**
Vin 1 0 AC 1
.AC LIN 500 3E6 6E6
**
.PARAM C1 = 190pF                    ; original value
*
.STEP PARAM C1 180pF 200pF 5pF   ; vary C1 (180-200pF)
*                                      in 5pF steps
**
R1 1 2 75
R2 3 0 75
C1 2 3 {C1}
L1 2 3 6.8 uH
*
.PROBE V(3)
*
.END
```

"loops" in conventional computer programming. To illustrate the use of the .STEP statement, consider the following example.

Example 5.3

Consider a requirement for a PAL colour subcarrier notch filter (4.43MHz). The required filter is shown in Figure 5.5.

Assume that the effect of varying C1 from 180 to 200pF on the filter response is to be examined. The PSpice input file of this circuit is given in Listing 5.3. This shows that the variation of C1 has been described using a combination of .PARAM and .STEP commands. PSpice will perform five analyses starting with C1 = 180pF and finishing with

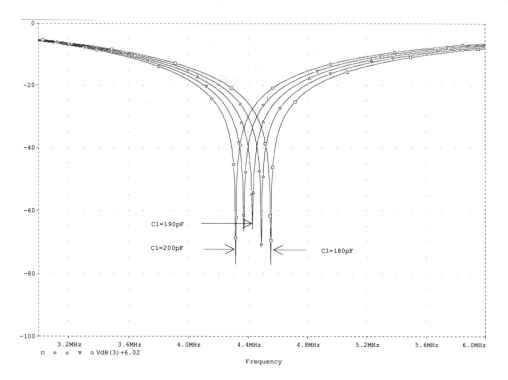

Figure 5.6   Frequency response of Example 5.3.

C1 = 200pF in steps of 5pF, while keeping the value of L1 unchanged during the various analyses.

Figure 5.6 shows the simulated filter response, which shows that notch frequency decreases as C1 increases, as expected.

5.4   Chapter summary

This chapter has shown, through worked examples, how PSpice can improve efficiency in circuit simulation by the use of subcircuits and step parameter statements. This was demonstrated through worked examples.

Reference

1. Al-Hashimi, B.M., "Building Bricks into Brick Wall Filters", *Electronics World + Wireless World*, June 1992, pp. 461-464.

chapter six

Analogue device models and libraries

The correlation between the simulated and the practical results depends on the accuracy of the component models used. So far, components have been assumed ideal. This chapter discusses the component models available in PSpice. There are two main types of models, passive and active. To give the user insight into how "good" or "bad" the simulation is, a comparison between the predicted and the practical results will be presented in selected instances.

6.1   Passive components

PSpice provides two methods for modelling components, the choice of which is determined by how the model is characterised. If the component is to be expressed in terms of an equivalent circuit, then the .SUBCKT command is the preferred method. If the component can be specified in terms of its parameters, then the .MODEL command is likely to be more appropriate. Table 6.1 shows the passive component models available in PSpice.

   PSpice has basic models for all passive components, and the "simpler" components (e.g., R, C, and L) have only .MODEL statements. For some of the more complex components such as crystals and transformers, PSpice has a library of models and the user must select the most appropriate model from the required library. In this section, the modelling of resistors, capacitors, and inductors is considered. (Model libraries are part of the PSpice simulation package and are written in plain text, which can be easily viewed or modified.)

6.1.1   Resistor model

So far, resistors have been described in PSpice using the basic description statement. This statement expresses a passive component in terms of component name, nodal connections, and component value. To model

Table 6.1   PSpice Passive Component Models

Description	PSpice name	.MODEL or .SUBCKT	Library
Capacitor	C	.MODEL	No
Inductor	L	.MODEL	No
Resistor	R	.MODEL	No
Quartz crystal	X	.SUBCKT	XTAL .LIB
Transmission line	T	See Appendix E	TLINE .LIB
Transformer core	K	.MODEL	MAGNETIC.LIB
Switch	S and W	See Appendix D	No

a resistor, two statements are required. The first is a description state-ment, and the second is a .MODEL statement. The description state-ment is

$$R<name> <node\ 1> <node\ 2> [model\ name] <value>$$

This shows that the basic resistor description statement now has an optional [*model name*]. This name, usually the descriptive name of the resistor, can begin with any character and can be up to eight characters long. The basic form of the .MODEL statement is

$$.MODEL <model\ name> RES [model\ parameters]$$

where the parameter <model name> is the name given to the resistor in the description statement. The parameter RES is the PSpice model symbol for the resistor. Table 6.2 shows the optional resistor model parameters.

This shows that it is possible to define temperature-dependent resistors. This is achieved by specifying the temperature coefficient(s) in the .MODEL statement. For example,

```
R1 1 2 RMOD 1K
R2 2 0 RMOD 2K
.MODEL RMOD RES (R = 1 TC1 = 0.00020)
```

Table 6.2   Resistor Model Parameters

Model parameter	Description	Default	Unit
R	Resistance multiplier	1	—
TC1	Linear temperature coefficient	0	$°C^{-1}$
TC2	Quadratic temperature coefficient	0	$°C^{-2}$
TCE	Exponential temperature coefficient	0	$\%/°C$

This describes two resistors (R1 and R2) with a model name, RMOD, which has been chosen arbitrarily. The .MODEL statement assumes that both resistors have a linear temperature coefficients of +200ppm/ °C. The other temperature coefficient of the resistor model has been assumed to be 0. The resistance multiplier, R, in the .MODEL statement has been set to 1 so that both resistors use the same resistor model. A fully worked example on the simulation of resistors with temperature coefficient is given in Chapter 7, Example 7.6.2.

6.1.2   Capacitor model

Capacitors are modelled in PSpice in a fashion similar to modelling of resistors. The capacitor modelling requires two statements, a description and a .MODEL statement. The description statement is

C<name> <node 1> <node 2> [model name] <value>

and the .MODEL statement is

.MODEL <model name> CAP [model parameters]

The parameter *CAP* is the PSpice model symbol for the capacitor. The capacitor model parameters are given in Table 6.3.

This shows that it is possible to define temperature- and voltage-dependent capacitors. This is achieved by specifying the temperature and voltage coefficients in the .MODEL statement. For example,

C1 5 6 CMOD 1μF
.MODEL CMOD CAP (C = 1 VC1 = 0.0001 TC1 = –0.000050)

This describes a capacitor with a linear voltage coefficient of 0.0001 and a temperature coefficient of –50ppm/°C.

Table 6.3   Capacitor Model Parameters

Model parameter	Description	Default	Unit
C	Capacitance multiplier	1	—
TC1	Linear temperature coefficient	0	$°C^{-1}$
TC2	Quadratic temperature coefficient	0	$°C^{-2}$
VC1	Linear voltage coefficient	0	V^{-1}
VC2	Quadratic voltage coefficient	0	V^{-2}

Table 6.4 Inductor Model Parameters

Model parameter	Description	Default	Unit
L	Inductance multiplier	1	—
TC1	Linear temperature coefficient	0	$°C^{-1}$
TC2	Quadratic temperature coefficient	0	$°C^{-2}$
IL1	Linear current coefficient	0	A^{-1}
IL2	Quadratic current coefficient	0	A^{-2}

6.1.3   Inductor model

Inductors are modelled similarly to modelling of resistors and capacitors. Two statements are required, a description and .MODEL statement. The description statement is

> L<name> <node 1> <node 2> [model name] <value>

and the .MODEL statement is

> .MODEL <model name> IND [model parameters]

where the parameter IND is the PSpice model symbol for the inductor. The inductor model parameters are given in Table 6.4.

This shows that it is possible to define temperature- and current-dependent inductors. This is achieved by specifying the temperature and current coefficients in the .MODEL statement. It should be noted that PSpice also allows the user to specify passive component (R, C, and L) tolerances using the .MODEL statement, and this is discussed in detail in Chapter 7.

Example 6.1
To check the accuracy of the passive component models, consider the circuit shown in Figure 6.1. It is a seventh-order lowpass elliptic filter

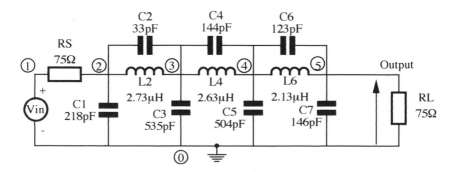

Figure 6.1   Ideal LC lowpass filter.

Listing 6.1a   PSpice Input File of Example 6.1

```
Elliptic lowpass filter (Figure 6.1)
**
Vin 1 0 AC 1
.AC LIN 500 0.1E6 30E6
*
RS 1 2 75
RL 5 0 75
*
C1 2 0 218pF
L2 2 3 2.73uH
C2 2 3 33pF
*
C3 3 0 535pF
L4 3 4 2.63uH
C4 3 4 144pF
*
C5 4 0 504pF
L6 4 5 2.13uH
C6 4 5 123pF
C7 5 0 146pF
*
.PROBE V(5)
*
.END
```

that has been designed to have a 5MHz passband edge, to provide >50dB attenuation at 8MHz, and to operate with 75Ω source and load impedance.

The PSpice input file of the filter is given in Listing 6.1a. The simulated frequency response of the filter is shown in Figure 6.2, and the measured filter response is shown in Figure 6.3. This shows that there is good agreement between the simulated and the practical filter response in terms of general filter shape and notch positions. The filter was built using TOKO coils (type 5P) and ceramic chip capacitors.

If, however, the filter passband is examined in detail, PSpice predicts an amplitude loss of less than 0.01dB (Figure 6.4), while the practical filter exhibits a loss of 0.8dB (Figure 6.5). This is a large relative error, and it is caused mainly by an inadequate model of the inductors. The built-in inductor model assumes infinite Q, but practical inductors always have losses associated with them.

A lossy inductor may be modelled using the simple equivalent circuit[1] shown in Figure 6.6, where R is given by

$$R = \frac{\omega L}{Q}$$

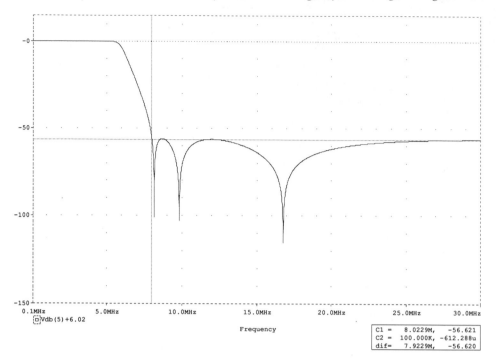

Figure 6.2   Simulated frequency response of circuit (Figure 6.1).

where ω is the frequency of interest, R is the inductor resistance, and Q is the quality factor of the inductor.

This lossy inductor equivalent circuit must be included in the filter simulation to improve the correlation between the simulated and the practical results. The Q of an inductor varies with frequency, and to provide a simplified inductor model, an average value of Q is taken from the published data sheet. For this inductor, Q is typically 40. The value of R can now be calculated from the above equation, assuming that ω=2πF. For simplicity, the frequency variable, F, is assumed to be equal to the filter passband edge, and in this case it is F=5MHz. Figure 6.7 shows the modelled filter using the calculated values of R.

The PSpice input file of this filter is given in Listing 6.1b, and the simulated frequency response of the modelled filter is shown in Figure 6.8, which compares favourably with that of the practical filter (Figure 6.5) in terms of the amplitude loss in the passband.

In summary, the existing PSpice passive component models (R, C, and L) allow the user to describe the following effects:

1. Resistor: temperature
2. Capacitor: temperature and variation of capacitor value with voltage

REF LEVEL   /DIV   MARKER 8 023 500.000Hz
−0.290dB   15.000dB   MAG (UDF)   −57.562dB

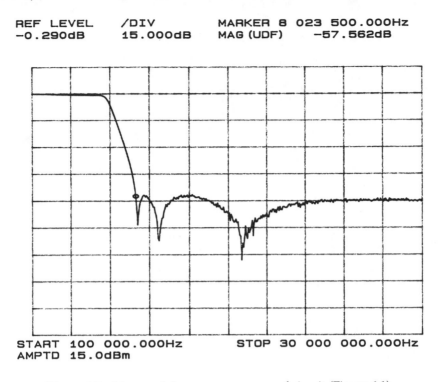

START 100 000.000Hz   STOP 30 000 000.000Hz
AMPTD 15.0dBm

Figure 6.3   Measured frequency response of circuit (Figure 6.1).

3. Inductor: temperature and variation of inductor value with current

Other effects such as frequency of operation and parasitic are not included which may be important depending on the application. Reference 15 discusses in detail passive component models that include these effects.

6.2   Semiconductor models

There are two main types of analogue semiconductor models, components and devices. Semiconductor component models include diodes, rectifiers, transistors, and thyristors. Semiconductor device models include op-amps, comparators, and regulators. PSpice also has models for opto-isolators and switch-mode regulators. Most PSpice semiconductor models are developed using the program *Parts*[2], which is part of the complete PSpice simulation package. This program allows the user to develop models for most active devices by following the program menu prompts. The data required by *Parts* are largely obtained from manufacturer's data sheets, although sometimes it may be necessary to provide actual measured results for certain parameters. *Parts* generates

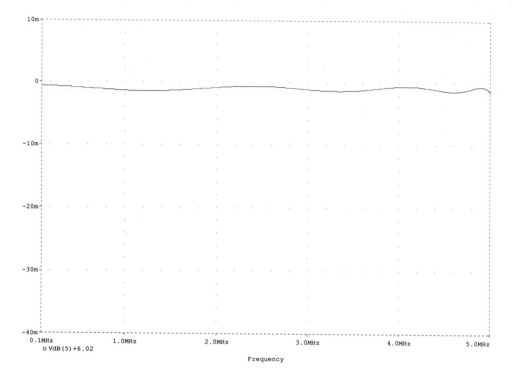

Figure 6.4 Simulated frequency response of circuit (Figure 6.1) over the frequency range of 0.1 to 5MHz.

models in the form of a .MODEL or a .SUBCKT definition that is recognised by PSpice. *Parts* is not discussed here, since for most users the PSpice libraries provide sufficient models. However, if it is necessary to develop a model for a component or device that is not part of the library, then the user is directed to Reference 2, which discusses the running of *Parts*. Note that *Parts* supports only diodes, transistors (bipolar, JFET, and power MOSFET), op-amps, comparators, and non-linear transformer magnetic cores. In this chapter only diode, bipolar transistor, and op-amp models will be discussed.

6.3   Diode model

Chapter 2 (Section 2.3.1) has discussed how diodes are described in PSpice. Two statements are required, a description and .MODEL statement. The description statement of the diode is

$$D<name> <NA> <NK> <model\ name>$$

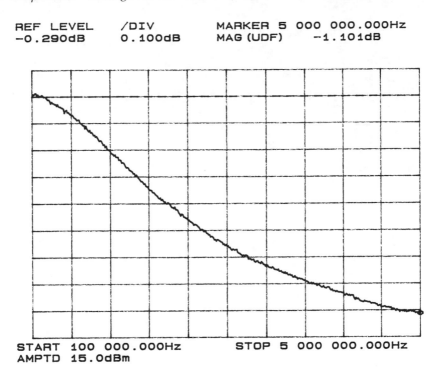

START 100 000.000Hz   STOP 5 000 000.000Hz
AMPTD 15.0dBm

Figure 6.5 Measured frequency response of circuit (Figure 6.1) over the frequency range of 0.1 to 5MHz.

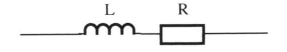

Figure 6.6 Simple equivalent circuit of a lossy inductor.

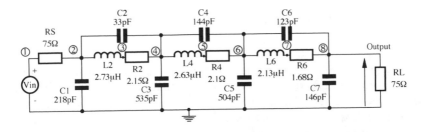

Figure 6.7 Lossy LC lowpass filter.

Listing 6.1b   PSpice Input File of Example 6.1

```
Modelled elliptic lowpass filter (Figure 6.7)
*
Vin 1 0 AC 1
.AC LIN 500 0.1E6 30E6
*
RS 1 2 75
RL 8 0 75
*
C1 2 0 218pF
L2 2 3 2.73uH
R2 3 4 2.15              ; L2 resistance, Q = 40, F = 5MHz
C2 4 0 33pF
*
C3 4 0 535pF
L4 4 5 2.63uH
R4 5 6 2.1               ; L4 resistance, Q = 40, F = 5MHz
C4 4 6 144pF
*
C5 6 0 504pF
L6 6 7 2.13uH
R6 7 8 1.68             ; L6 resistance, Q = 40, F = 5MHz
C6 6 8 123pF
C7 8 0 146pF
*
.PROBE V(8)
*
.END
```

where <NA> and <NK> are the anode and the cathode of the diode, respectively, and <*model name*> is usually a descriptive name of the diode. This name, which is chosen arbitrarily by the user, can begin with any character and can be up to eight characters long. Examples are diode1 and type1. The .MODEL statement of a diode is

.MODEL <model name> D [model parameters]

The parameter <*model name*> is the name given to the diode in the description statement, and *D* is the PSpice symbol for the diode. PSpice has a detailed and complex diode model that can be used to simulate many DC and AC characteristics. The model has 25 parameters, as shown in Table 2.4 of Chapter 2. To model a diode, the user must specify values of the model parameters. If no values are specified, PSpice will use the default values shown in Table 2.4. For simple circuit analysis, it is generally sufficient to allow model parameters to have

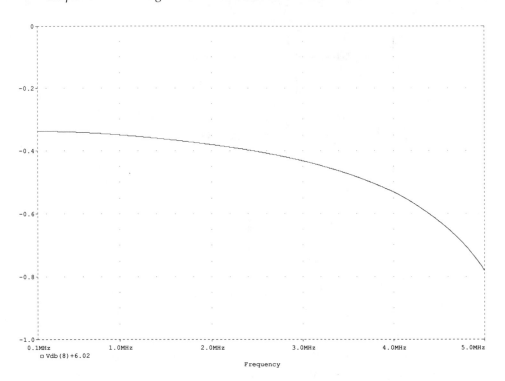

Figure 6.8   Simulated frequency response of lossy LC lowpass filter (Figure 6.7).

their default values so that an ideal diode performance is obtained. To illustrate this point, consider the following example.

Example 6.2

Figure 6.9 shows a half-wave rectifier with reservoir capacitor, and the PSpice input file of this circuit is given in Listing 6.2. The diode, D1, is connected between nodes 1 and 2, with the anode at node 1. The model name of this diode, TYPE1, has been chosen arbitrarily. The diode model parameters have been set to their default values, since no model parameters were specified in the .MODEL statement of the diode. Figure 6.10 shows the simulated response of the rectifier, assuming a sine wave input signal of 60Hz and 40V peak-to-peak amplitude.

There are situations, however, that would require accurate diode models. In this case, the user must have a good understanding of the model and its parameters. For a more detailed discussion of *SPICE* semiconductor models, see References 3 to 6. PSpice, however, has more than 400 models of commercially available diodes. Theses models are in the DIODE .LIB library as shown in Table 6.5.

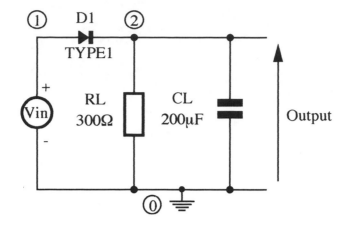

Figure 6.9 Circuit of Example 6.2.

The DIODE .LIB library consists of small-signal, power, Zener, voltage-variable capacitance diodes and rectifiers. The EUROPE .LIB library includes models of diodes that use the European numbering system, unlike the DIODE .LIB library, which uses the American numbering system. To use a library model, the user must first describe the diode, using the diode description statement given earlier, and the *<model name>* must now be the name of the diode selected from the library. Also, the appropriate library name must be specified in the input file using a .LIB statement so that the required diode model is chosen. This means that the .MODEL statement is not required when a library model is used. To demonstrate this, consider the following example.

Listing 6.2 PSpice Input File of Example 6.2

```
Half-wave rectifier circuit (Figure 6.9)
*
Vin 1 0 SIN (0V 20V 60Hz)
.TRAN 0.02ms 50ms 0ms 0.02ms
*
D1 1 2 TYPE1           ; diode connections
.MODEL TYPE1 D         ; user-defined diode model
RL 2 0 300
CL 2 0 200uF
*
.PROBE V(1), V(2)
.END
```

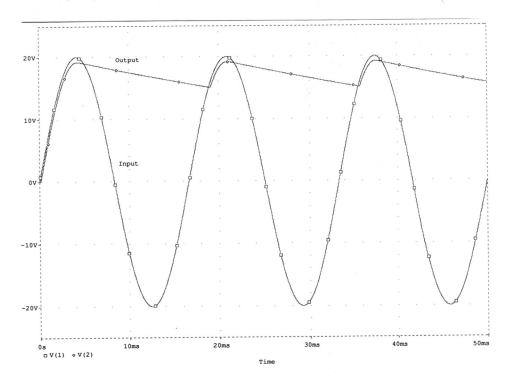

Figure 6.10 Transient response of Example 6.2.

Example 6.3

To simulate the half-rectifier circuit shown in Figure 6.9, use the commercially available power diode 1N4002. The PSpice input file of this circuit is given in Listing 6.3. Note than the diode name is D1N4002 rather than 1N4002. This is because of a PSpice requirement that the model name must begin with a letter. As a result all library diode models are prefixed with the letter D. The .LIB statement directs PSpice to the location of the diode "D1N4002", and in this case it is "C:\diode.lib". This assumes that the diode model library is on drive C of the PC. The D1N4002 model held in DIODE .LIB PSpice library is

.MODEL D1N4002 D(IS=14E-9, N=1.98, IKF=94.81, BV=150,
+ IBV=10E-6, RS=0.0339, TT=5.7E-6, CJO=25.89E-12, VJ=0.3245,
+ M=0.44)

Table 6.5 Diode PSpice Models

Description	PSpice name	.MODEL or SUBCKT	Library
Diode	D	.MODEL	DIODE.LIB
			EUROPE.LIB

Listing 6.3   PSpice Input File of Example 6.3

```
Half-wave rectifier circuit  (Figure 6.9)
*
Vin 1  0  SIN  (0V 20V 60Hz)
.TRAN 0.02ms 50ms 0ms 0.02ms
*
D1  1  2  D1N4002                    ;  1N4002 diode connections
.LIB  C:\diode.lib                   ;  diodes model library
*
RL  2  0  300
CL  2  0  200uF
*
.PROBE  V(1),  V(2)
*
.END
```

with the remaining diode model parameters set to their default values
(Table 2.4). The simulated response of this rectifier is effectively the
same as that in Figure 6.10. In this case the ideal and actual diode
models have produced the same results. The user must decide when to
use the ideal or actual model, depending on the application.

6.4   *Bipolar transistor model*

Chapter 2 (Section 2.3.2) discussed how bipolar transistors are described
in PSpice. Two statements are required, a description and a .MODEL
statement. The description statement of the bipolar transistor is

$$Q<name> <NC> <NB> <NE> <model\ name>$$

where *<NC>*, *<NB>*, and *<NE>* are the collector, base, and emitter
nodes, respectively. The parameter *<model name>* is usually a descriptive name of the diode. This name, which is chosen arbitrarily by the
user, can begin with any character and can be up to eight characters
long. Examples are tran and Q1. The .MODEL statement of a bipolar
transistor is

$$.MODEL <model\ name> <type> [model\ parameters]$$

The parameter *<model name>* is the name given to the transistor in
the description statement. The parameter *<type>* determines the transistor type (NPN or PNP). PSpice has a detailed and complex transistor

Table 6.6 Transistor PSpice Models

Description	PSpice name	.MODEL or .SUBCKT	Library
Bipolar transistor	Q	.MODEL	BIPOLAR.LIB
			EUROPE.LIB
			PWRBJT.LIB
JFET	J	.MODEL	JFTE.LIB
MOSFET	M	.MODEL	PWRMOS.LIB
GaAsFET	B	.MODEL	No

model that can be used to model many DC and AC characteristics. The model has more than 40 parameters, as shown in Table 2.5 of Chapter 2. To model a bipolar transistor, the user must specify values for the model parameters. If no values are specified, PSpice will use the default values shown in Table 2.5. For simple circuit analysis, it is generally sufficient to allow model parameters to have their default values so that an ideal transistor performance is obtained. See Examples 2.7.1 and 2.7.2 of Chapter 2 for the simulation of ideal bipolar transistors. There are situations, however, that would require accurate transistor models, such as the development and design of a new integrated circuit. In this case, the user must have a good understanding of the model and its parameters. For a detailed discussion of *SPICE* semiconductor models, see References 3 to 6. PSpice offers the designer a wide choice of accurate models of commercially available transistors. Table 6.6 shows the transistor models available. Again, the EUROPE .LIB includes models of bipolar transistors that use the European numbering system.

To use a transistor library model, the user must first describe the device using a description statement, and the <*model name*> must now be the name of the transistor selected from the library. Also, the appropriate library name must be specified in the input file using a .LIB statement so that the required transistor model is chosen.

Example 6.4
To check the accuracy of the PSpice diode and bipolar transistor models, consider the simulation of the circuit in Figure 6.11. It is a low-power video amplifier, with D1 providing a clamping action. The PSpice input file is given in Listing 6.4. The diode (1N4148) model is specified using the first .LIB statement. The bipolar transistor models are specified using the second .LIB statement. The simulated amplifier frequency response is shown in Figure 6.12, which compares reasonably well with the practical amplifier response shown in Figure 6.13. Table 6.7 summarises the comparison between the simulated and the practical results.

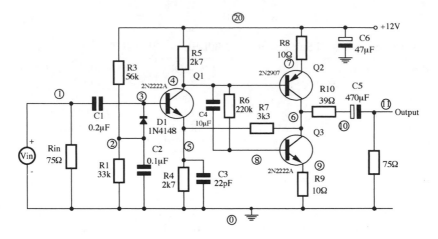

Figure 6.11   Circuit of Example 6.4.

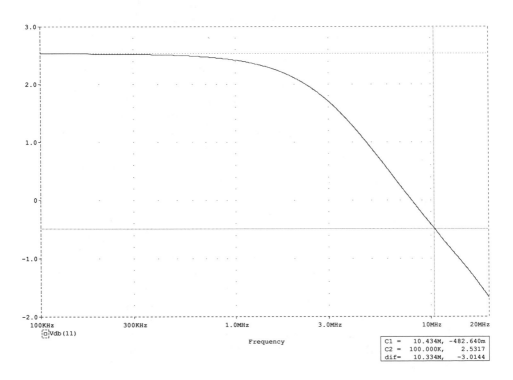

Figure 6.12   Simulated frequency response of circuit (Figure 6.11).

Listing 6.4 PSpice Input File of Example 6.4

```
Transistor video amplifier (Figure 6.11)
*
Vin 1 0 AC 1
.AC LIN 800 0.1E6 25E6
VCC 20 0 +12V                          ; 12V power supply
*
RIN 1 0 75
C1 1 3 0.2uF
R1 2 0 33K
R3 2 20 56K
C2 2 0 0.1uF
R4 5 0 2K7
C3 5 0 22pF
R5 4 20 2K7
R6 4 8 22K
C4 4 8 10uF
R7 5 6 3K3
R8 7 20 10
R9 9 0 10
R10 6 10 39
C5 10 11 470uF
C6 20 0 47uF
RL 11 0 75
*
D1 2 3 D1N4148                         ; diode description
.LIB C:\diode.lib                      ; diodes model library
Q1 4 3 5 Q2N2222A                      ; NPN description
Q2 6 4 7 Q2N2907                       ; PNP description
Q3 6 8 9 Q2N2222A                      ; NPN description
.LIB C:\bipolar.lib                    ; transistor model library
*
.PROBE V(11)
*
.END
```

Table 6.7 Comparison Between Simulation and Practical Results

Parameter	Simulation	Practical
Gain at 100kHz	2.53dB	1.37dB
−3dB Bandwidth	10.33MHz	9.35MHz
Quiescent current	10.4mA	10mA

REF LEVEL   /DIV   OFFSET 9 347 505.400Hz
1.370dB   0.500dB   MAG (UDF)   -3.040dB

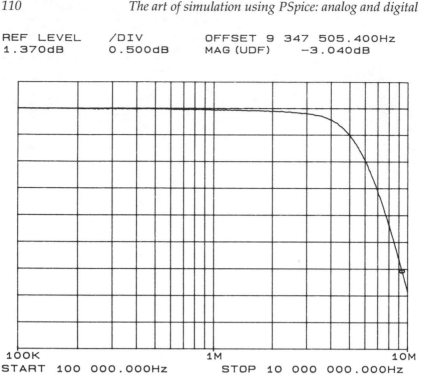

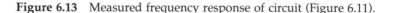

100K   1M   10M
START 100 000.000Hz   STOP 10 000 000.000Hz

Figure 6.13   Measured frequency response of circuit (Figure 6.11).

6.5   *Operational amplifier model*

Chapter 4 (Section 4.1.1) showed how ideal op-amps are described using voltage-controlled voltage sources (VCVS). Chapter 4 also illustrated how the op-amp gain-frequency roll-off characteristics can be modelled using a simple RC network; see Example 4.1, Listing 4.1b. To model other op-amp parameters such as slew rate, bias current, and maximum output voltage swing, a more complex op-amp model is needed. Figure 6.14 shows the internal circuitry of the 741 op-amp. Based on this circuit, a transistor-level op-amp model can be defined in terms of a subcircuit and used when required.

Using this model, however, would require a large amount of simulation time. Although computing power has increased, these advances have been overshadowed by the need to simulate more complex circuits. In order to reduce simulation time, a compressed version of the transistor-level op-amp model has been developed. This compressed op-amp model is known as a "macromodel". Figure 6.15 represents the

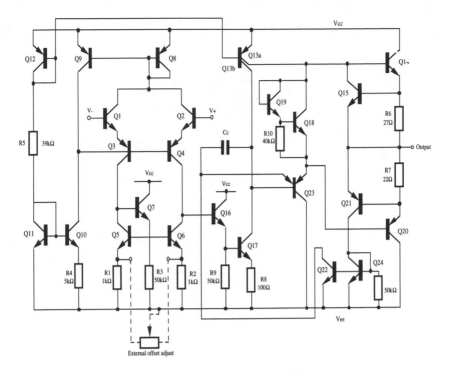

Figure 6.14 Internal circuitry of a 741 op-amp.

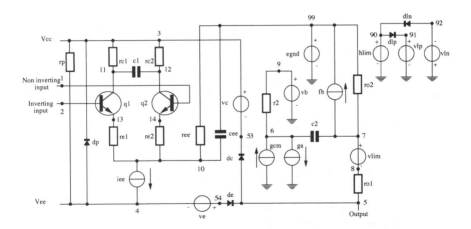

Figure 6.15 The 741 op-amp macromodel.

Table 6.8 Op-amp Macromodel Components

Component	Description
c1	Phase-control capacitor
c2	Compensation capacitor
cee	Slew-rate-limiting capacitor
dp	Substrate junction
egnd	Voltage-controlled voltage source
fb	Output device
ga	Interstage transconductance
gcm	Common-mode transconductance
iee	Input-stage current
hlim	Voltage-limiting device
q1,q2	Input transistors
r2	Interstage resistance
rc1,rc2	Input-stage load resistance
re1,re2	Input-stage emitter resistance
ree	Input-stage current source output resistance
ro1	Output resistor No. 1
ro2	Output resistor No. 2
rp	Power dissipation
vb	Independent voltage source
vc,dc	Output offset limiter (to Vcc)
ve,de	Output offset limiter (to Vee)
vlim	Output current-limiting sensor
v1n,d1n	Negative supply limit
v1p,d1p	Positive supply limit

741 op-amp macromodel, which shows that the nodes and component count have been reduced considerably when compared to that of Figure 6.14. These macromodels yield a large reduction in simulation time. Table 6.8 lists and describes the macromodel components. This op-amp macromodel was first described by Boyle et al.[7], and hence, it is often referred to as the Boyle model.

In order to determine the macromodel values, it is necessary to use *Parts*. This software allows the user to enter the key op-amp parameters (such as open loop gain and slew rate) from the manufacturer's data sheet and then calculates the op-amp macromodel values. Box 6.1 shows the UA741 macromodel subcircuit[8] description. The op-amp macromodel nodal notation is given in Figure 6.16, which shows that node 1 is the noninverting input, node 2 is the inverting input, nodes 3 and 4 are the positive and negative power supplies, respectively, and node 5 is the op-amp output.

PSpice has a wide choice of commercially available op-amp models, as shown in Table 6.9. These models were generated by the semi-

Box 6.1   Texas Instruments UA741 Op-amp Macromodel Subcircuit
Description

```
* Connections   noninverting input
*               | inverting input
*               | | positive power supply
*               | | | negative power supply
*               | | | | output
*               | | | | |
*               | | | | |
*               | | | | |
.SUBCKT UA741  1  2  3  4  5
*
c1 11 12 4.664E-12
c2 6 7 2.00E-12
dc 5 53 DX
de 54 5 DX
d1p 90 91 DX
d1n 92 90 DX
dp 4 3 DX
*
*
egnd 99 0 POLY (2) (3,0) (4,0) 0.5.5
fb 7 99 POLY(5) VB VC VE VLP VLN 0 10.61E6 -10E6 10E6 10E6 -10E6
ga 6 0 11 12 137.7E-6
gcm 0 6 10 99 2.574E-9
iee 10 4 DC 10.16E-6
hlim 90 0 Vlim 1K
*
*
q1 11 2 13 QX
q2 12 1 14 QX
r2 6 9 100.0E3
rc1 3 11 7.957E3
rc2 3 12 7.957E3
re1 13 10 2.740E3
re2 14 10 2.740E3
ree 10 99 19.69E6
ro1 8 5 15
ro2 7 99 15
rp 3 4 18.11E3
vb 9 0 DC 0
vc 3 53 DC 2.600
ve 54 4 DC 2.600
vlim 7 8 DC 0
vlp 91 0 DC 25
vln 0 92 DC 25
*
*
.MODEL DX D(IS=800.0E-18)
.MODEL QX NPN(IS=800.0E-18 BF=62.50)
.END
```

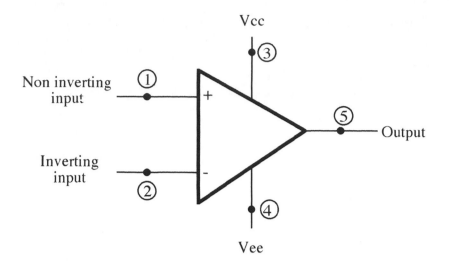

Figure 6.16   Op-amp subcircuit nodal notation.

conductor manufacturers using the program *Parts*, apart from the LIN-EAR .LIB, which was developed by the PSpice vendor. This library consists of popular op-amp models.

Op-amp models developed using *Parts* are capable of modelling

1.   DC input impedance and bias current
2.   Differential and common mode gain
3.   Open loop gain-frequency characteristic
4.   Slew rate
5.   Maximum output voltage swing
6.   Output impedance
7.   DC quiescent power

Table 6.9   PSpice Op-amp Models

Description	PSpice name	.MODEL or .SUBCKT	Library
Op-amps	X	.SUBCKT	ANLG_DEV.LIB
			BURR_BRN.LIB
			COMLINR.LIB
			ELANTEC.LIB
			HARRIS.LIB
			LINEAR.LIB
			LIN_TECH.LIB
			NAT_SEMI.LIB
			TEX_INST.LIB

However, it does not model the following parameters:

1. Input offset voltage and offset bias current
2. Noise and distortion
3. Variations in performance with temperature

Some semiconductor manufacturers have improved the op-amp macromodels generated by *Parts* to the extent that they are capable of modelling input offset voltage and input offset current, as well as input voltage and input current noise. Also, the op-amp open loop gain-frequency characteristic can be described in terms of zeros as well as poles. This type of op-amp model is often referred to as the enhanced op-amp macromodel. For a detailed discussion on this model, see References 9 and 10. Different manufacturers employ different techniques to model some of these op-amp features[10].

6.5.1 Using op-amp macromodels

To use (or call) an op-amp library model, an X statement is required since the op-amp is described in terms of a subcircuit. The basic form of an op-amp X statement is

X<name> <+ node> <– node> <Vcc> <Vee> <output> <model name>

where *<+ node>* is the noninverting input node, and *<– node>* is the inverting input node of the op-amp. The parameter *<Vcc>* is the op-amp positive power supply, *<Vee>* is the negative power supply, and *<output>* is the op-amp output node. The parameter *<model name>* is the library PSpice model name of the op-amp. The required op-amp library must be specified in the input file using a .LIB statement. The following two examples illustrate various uses of the op-amp macromodel.

Example 6.5
To simulate the frequency response of the amplifier circuit shown in Figure 6.17, assume that the op-amp is the 741 manufactured by Texas Instruments.

The input file of the amplifier circuit is given in Listing 6.5. The op-amp has been described by an "X1" device with the op-amp noninverting input at node 1, the inverting input at node 2, the positive power supply at node 10, the negative power supply at node 11, and the output at node 3. It has been assumed that op-amp requires ±15V power supplies. It is important to specify the nodes of the op-amp subcircuit in the correct order. The model name of the 741 op-amp in the TEX_INST.LIB is UA741/301/TI. The library is referred to using the .LIB statement.

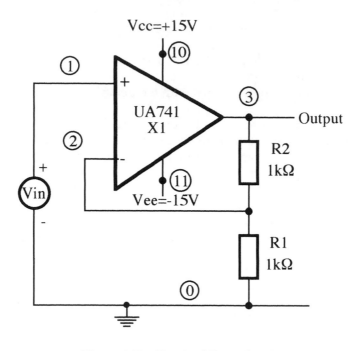

Figure 6.17   Circuit of Example 6.5.

Listing 6.5   PSpice Input File of Example 6.5

```
Amplifier Circuit  (Figure 6.17)
*
Vcc 10 0 DC +15V                    ; +15V power supply
Vee 11 0 DC -15V                    ; -15V power supply
*
Vin 1 0 AC 1
.AC LIN 800 1 10E6
R1 2 0 1K
R2 2 3 1K
*
* + - Vcc Vee output op-amp name    ; op-amp nodes order
*
X1 1 2 10 11 3 UA741/301/TI   ; UA741 op-amp description
.LIB C:\TEX_INST.LIB          ; Texas Instruments library
*
.PROBE V(3)
*
.END
```

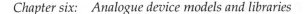

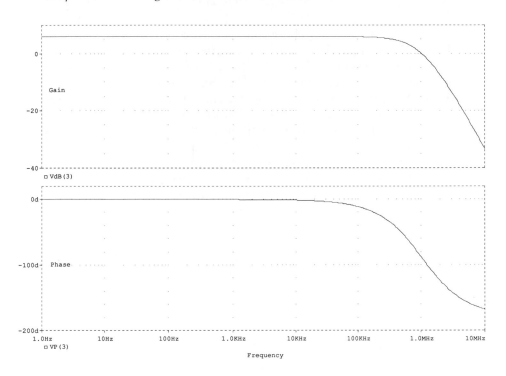

Figure 6.18 Frequency response (gain and phase) of Example 6.5.

Note that the numbers and the letters after the op-amp model name (UA741) are Texas_Instruments special notations.

Figure 6.18 shows the simulated frequency response (gain and phase) of the amplifier.

Example 6.6

To check the accuracy of the op-amp macromodels, consider the circuit in Figure 6.19. It is a ninth-order lowpass FDNR elliptic filter, which has been designed to have a 20kHz passband edge and provide >60dB attenuation at 22.22kHz. This circuit has been discussed previously in

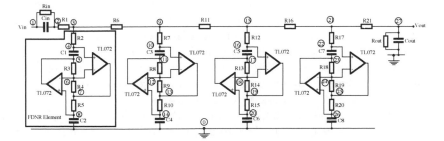

Figure 6.19 A 20kHz lowpass FDNR filter.

Chapter 5 and has been chosen here because of its complexity. In
Chapter 5 (Example 5.1) the FDNR elements of the filter have been
described in terms of a subcircuit called FDNR, and the op-amps were
assumed ideal (i.e., described using VCVS). Here, the FDNR elements
will still be described in terms of a subcircuit, but this time the op-amps
will be assumed "practical" (i.e., described using macromodels). In
practice, the op-amps of each FDNR element will be realised using a
dual op-amp package because close matching is required. In this case
the dual op-amp TL072 from Texas Instruments will be used. This
amplifier is a low-noise device with a unity-bandwidth product of
3MHz. The "practical" FDNR element (Figure 6.20) subcircuit description is

```
.SUBCKT FDNR_M 1 7 PARAMS: RA=1, R=1, C=1
*

Vcc 20 0 15V                        ; +15V power supply
Vee 21 0 -15V                       ; -15V power supply
*

RA 1 2 {RA}
CA 2 3 {C}
RB 3 4 {R}
RC 4 5 {R}
RD 5 6 {R}
CB 6 7 {C}
*

X1 2 4 20 21 5 TL072/301/TI         ; TI TL072 op-amp
X2 6 4 20 21 3 TL072/301/TI         ; TI TL072 op-amp
*

.ENDS FDNR_M                        ; end of subcircuit definition
```

It has been assumed that the FDNR subcircuit is called FDNR_M.

The lowpass FDNR filter in terms of subcircuits is shown in Figure
6.21 with the PSpice input file in Listing 6.6. The FDNR_M subcircuit is
included in a file called "circuits" as shown in the first .LIB statement
of Listing 6.6.

In a high-performance filter such as the one described in this example, the bandwidth is often quoted to be the –0.3dB point (unlike
simple filters as where the –3dB point is usually quoted). Assuming
ideal op-amp models, PSpice predicts a filter with 20kHz bandwidth
(Figure 6.22) designed. The op-amp macromodel simulation predicts a
bandwidth of 19.75kHz (Figure 6.23), while the practical filter produces
a bandwidth of 19.6kHz (Figure 6.24). The practical filter was built
using selected and trimmed capacitors and resistors so that the performance was mostly determined by the op-amps.

This example has demonstrated that both the ideal and macromodel
simulations produce results close to practical results, and a number of
filter parameters can be compared (see Table 6.10). Figure 6.25 shows

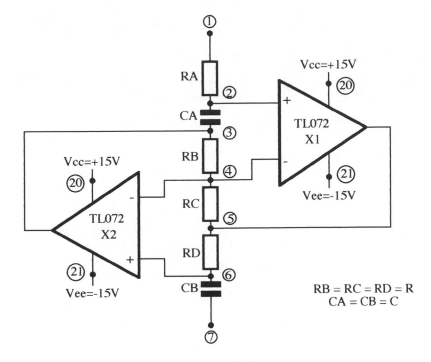

Figure 6.20 Practical FDNR element.

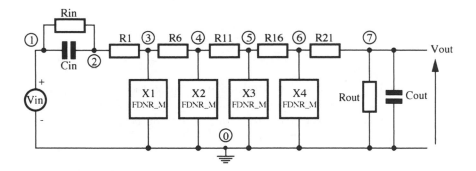

Figure 6.21 FDNR filter (Figure 6.19) with subcircuit notation.

the ideal overall filter response with the notch positions indicated. Note that Table 6.10 shows that the practical filter notches have been shifted to frequencies lower than the ideal. This has resulted in a sharper filter response.

In an attempt to improve the accuracy between the macromodel simulation and the practical result, a transistor-level op-amp model based on the TL072 circuit was developed. When this was tested in the

Listing 6.6   PSpice Input File of Example 6.6

```
20kHz FDNR Lowpass Filter (Figure 6.21)
*

.LIB A:\circuits          ; location of FDNR_M subcircuit
.LIB C:\TEX_INST.LIB      ; TI TL072 op-amps library
*

Vin 1 0 AC 1
.AC LIN 800 1 50K
*

Cin 1 2 10nF
Cout 7 0 10nF
Rin 1 2 1MEG
Rout 7 0 1MEG
*

R1 2 3 1.1k
X1 3 0 FDNR_M params: RA=117, R=968, C=10nF
*

R6 3 4 1.36k
X2 4 0 FDNR_M params: RA=0.735k, R=0.610k, C=10nF
*

R11 4 5 0.948k
X3 5 0 FDNR_M params: RA=1.01k, R=0.5k, C=10nF
*

R16 5 6 1.08k
X4 6 0 FDNR_M params: RA=0.441k, R=0.711k, C=10nF
*

R21 6 7 0.856k
**

.PROBE V(7)
*

.END
```

filter circuit, PSpice failed to produce an answer due to convergence error. This error is indicated by PSpice in the circuit output file. This example shows that the user must decide on the complexity of the model to achieve the required accuracy.

6.6   *Dealing with unstable circuits*

To determine theoretically the stability of a circuit, the poles and zeros of the circuit transfer function must be known. Unfortunately, PSpice does not provide such information. However, PSpice is capable of predicting the performance of conditionally stable circuits, provided the correct analysis type is chosen. If the frequency response (i.e., .AC statement) alone is used, the outcome may appear to be a stable circuit. In order to check for instability, it is essential that a transient analysis (i.e., .TRAN statement) is performed as well. The following two ex-

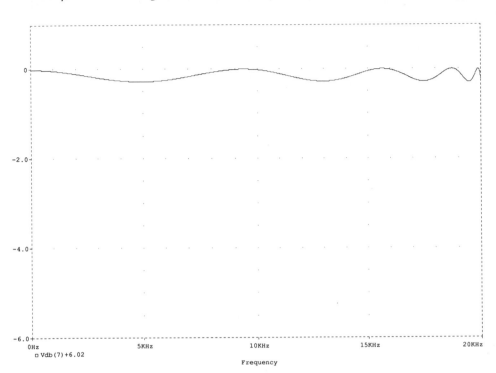

Figure 6.22 Simulated frequency response of circuit (Figure 6.21) using ideal op-amps.

amples will demonstrate the capability of PSpice in predicting instability problems in circuits.

Example 6.7

Check the stability of the amplifier circuit shown in Figure 6.26. As can be seen, this amplifier has a positive feedback. The input file of this circuit is given in Listing 6.7. PSpice allows the user to define several independent sources using one statement. For example, in Listing 6.7, the "Vin" source produces an AC signal of 1V between nodes 1 and 0, as well as a 5kHz sine wave of 5V amplitude between the same nodes. Performing a frequency response simulation gives the results of Figure 6.27. This shows that the amplifier has a gain of +2 (or 6dB) at low frequency, which is theoretically correct.

However, if transient analysis of this circuit is now performed, the results of Figure 6.28 are obtained, which confirms what would be likely to happen in practice (i.e., saturation). It has been assumed that sine wave waveform of 10V peak-to-peak amplitude and 1kHz frequency is used in the transient analysis.

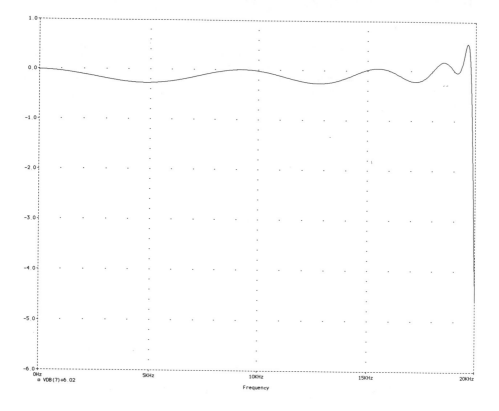

Figure 6.23 Simulated frequency response of circuit (Figure 6.21) using macromodel TL072 op-amps.

Example 6.8
The late 1980s brought a new family of op-amps onto the IC marketplace. These are based on current feedback[12,13] rather than voltage feedback, which is usually used in conventional op-amps, and this has resulted in devices with a much higher bandwidth. A wide variety of such op-amps are commercially available. One manufacturer of current feedback amplifiers (CFAs) is Elantec[14]. PSpice has an Elantec library of CFA models (see Table 6.9). With voltage feedback amplifiers (VFAs), the choice of the feedback resistor is almost arbitrary, provided the ratio of RF to R1 is correct to achieve the required gain (Figure 6.29). However, with CFAs the feedback resistor sets the amplifier bandwidth and the frequency response shape, as well as defining the gain. The data sheet of the EL2030 amplifier indicates that if the value of the feedback resistor, RF, in the amplifier circuit (Figure 6.29) is reduced below 200Ω, oscillation will occur.

REF LEVEL   /DIV   MARKER 20 000.000Hz
2.000dB   2.000dB   MAG (UDF)   -9.813dB

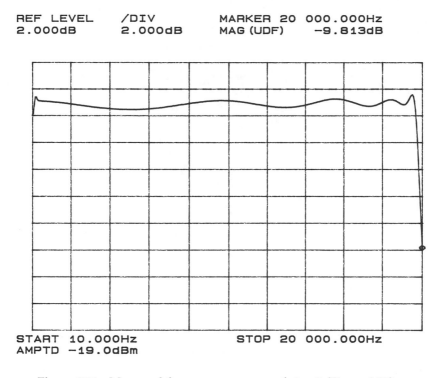

START 10.000Hz   STOP 20 000.000Hz
AMPTD -19.0dBm

Figure 6.24   Measured frequency response of circuit (Figure 6.21).

The PSpice input file of the amplifier circuit is given in Listing 6.8. Note that multiple independent sources are defined using one state-ment similar to that of Listing 6.7. The simulated frequency response of the amplifier for three different values of RF (150Ω, 250Ω, and 1kΩ) is given in Figure 6.30, and it has been assumed that R1=RF.

If a transient analysis of the amplifier circuit is now performed, the results of Figure 6.31 are obtained, which clearly indicate that the amplifier is unstable when RF=R1=150Ω (i.e., oscillation occurs). It has

Table 6.10   Comparison between Simulation and Practical Results

Parameter	Ideal	Macromodel	Practical
Bandwidth (–0.3dB point)	20kHz	19.75kHz	19.6kHz
Attenuation at 22.22kHz	63dB	66dB	69dB
1st Notch position (F1)	22.40kHz	22.02kHz	21.5kHz
2nd Notch position (F2)	23.76kHz	23.5kHz	23kHz
3rd Notch position (F3)	28.39kHz	27.81kHz	27.5kHz
4th Notch position (F4)	47.29kHz	44.83kHz	44.99kHz

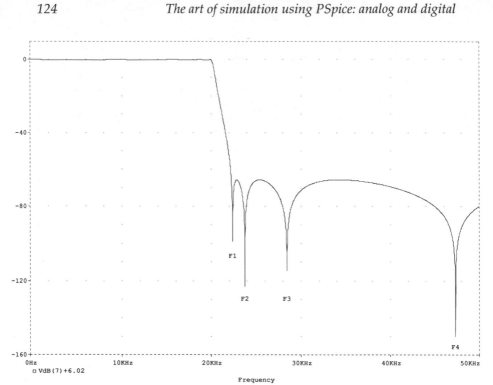

Figure 6.25   Overall ideal frequency response of circuit (Figure 6.21) with notch positions.

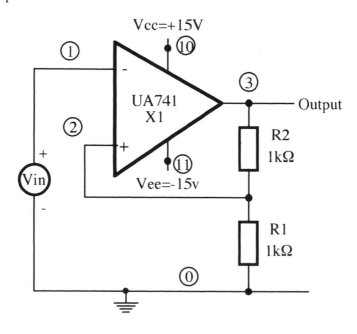

Figure 6.26   Circuit of Example 6.7.

Listing 6.7 PSpice Input File of Example 6.7

```
Positive Feedback Amplifier Circuit (Figure 6.26)
*
*
Vcc 10 0 DC +15V                  ; +15V power supply
Vee 11 0 DC -15V                  ; -15V power supply
*
Vin 1 0 AC 1 SIN (0V 5V 1kHz)    ; AC & transient i/p signals
.AC LIN 500 1 10E6
.TRAN 0.01ms 3ms 0ms 0.01ms
*
R1 2 0 1K
R2 2 3 1K
*
* + - Vcc Vee output model name ; op-amp nodes order
*
X1 2 1 10 11 3 UA741/301/TI   ; TI UA741 op-amp description
.LIB C:\TEX_INST.LIB       ; Texas Instruments op-amp library
*
.PROBE V(1), V(3)
*
.END
```

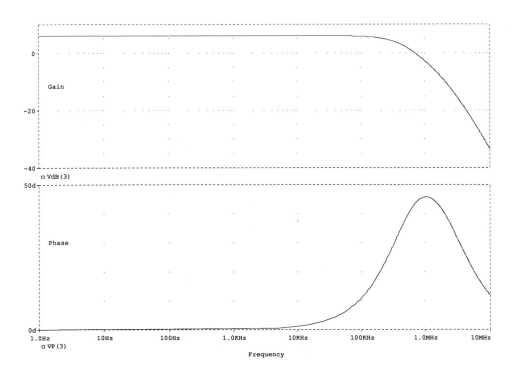

Figure 6.27 Frequency response (gain and phase) of circuit (Figure 6.26).

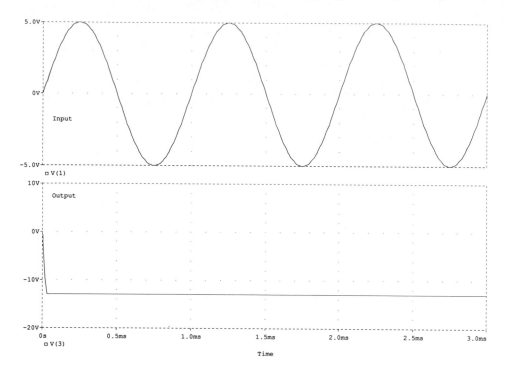

Figure 6.28　Transient response of circuit (Figure 6.26).

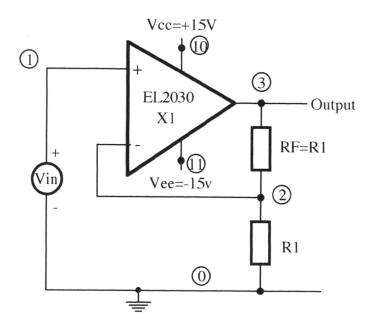

Figure 6.29　Circuit of Example 6.8.

Listing 6.8   PSpice Input File of Example 6.8

```
Current Feedback Amplifier (Figure 6.29)
*
Vin 1 0 AC 1 PULSE (0V 1V 0s 10ns 10ns 0.1us 0.2us)
.AC LIN 600 0.1E6 200E6
.TRAN 0.01us 0.4us 0s 0.01us
*
Vcc 10 0 +15V                          ; +15V power supply
Vee 11 0 -15V                          ; -15V power supply
*
R1 2 0 150
RF 2 3 150
*
X1 1 2 10 11 3 EL2030/EL               ; EL2030 op-amp connection
.LIB C:\ELANTEC.LIB                    ; Elantec op-amp library
*
.PROBE V(1), V(3)
*
.END
```

been assumed that a 1MHz square waveform has been applied to the circuit input.

PSpice allows the user to combine a number of input files so that multiple analyses of the same circuit can be performed. Listing 6.8 was copied three times (for three different values of RF=R1) to generate a single input file. Note that each of the three files starts with a comment line and finishes with an .END statement. To plot the simulation result of each analysis as shown in Figure 6.31, the symbol "@" must be used to specify the required waveform. For example, V(3)@2 means the amplifier output voltage of the second analysis. To simplify the comparison between the input and output waveforms of the amplifier, the output voltage, V(3), has been multiplied by 0.5, as shown in Figure 6.31. It should be noted that if no @ symbol is specified when a number of analyses have been performed, *Probe* will plot the results of all analyses, as shown in Figure 6.30.

6.7   Chapter summary

- PSpice has good component modelling capabilities, with comprehensive libraries available.
- Most PSpice models are developed using the program *Parts*. This program requires input data usually obtained from the manufacturer's data sheets.

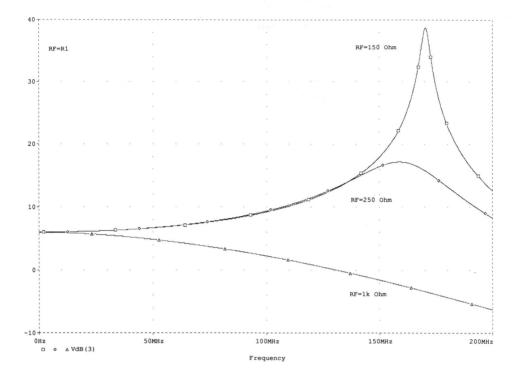

Figure 6.30 Frequency response of circuit (Figure 6.29) for different values of RF.

- *Parts* generates models in the form of a .MODEL or a .SUBCKT definition.
- Close agreement between simulated and practical results is generally possible.
- The user must decide on the level of model complexity to achieve the required accuracy.

References

1. Sedra, A.S., & Brackett, P.O., *Filter Theory and Design: Active and Passive*, Pitman Publishing Limited, London 1979, pp. 282.
2. MicroSim Corporation, *Circuit Analysis — User's Guide* (The Design Center, Version 5.3, Irvine, CA, January 1993.
3. Antognetti, P., & Massobrio, G., *Semiconductor Device Modelling with SPICE*, McGraw-Hill, New York, 1988.
4. Gummel, H.K., & Poon, H.C., "An Integral Charge Control Model for Bipolar Transistors", *Bell System Technical Journal*, Vol. 49, May/June 1970, pp. 827-852.
5. Schichman, H., & Hodges, D.A., "Modelling and Simulation of Insulated-Gate Field-Effect Transistor Switching Circuits", *IEEE Journal of Solid-State Circuits*, Vol. SC-3, September 1968, pp. 285-289.

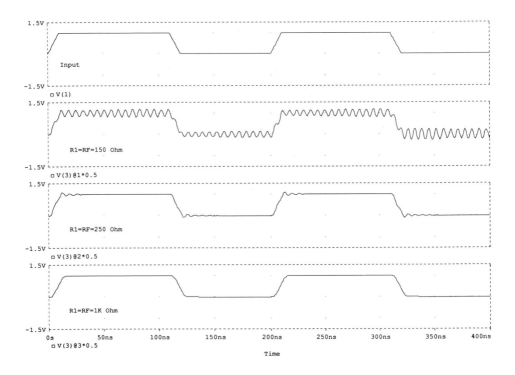

Figure 6.31 Transient response of circuit (Figure 6.29) for different values of RF.

6. Sheu, B.J., Scharfeter, D.L., Ko, P., & Jeng, M., "BSIM: Berkeley Short-Channel IGFET Model for MOS Transistors", *IEEE Journal of Solid-State Circuits*, Vol. 22, No. 4, 1987, pp. 558-566.

7. Boyle, G.R., Cohn, B.M., Pederson, D., & Solomon, J.E., "Macromodelling of Integrated Circuit Operational Amplifiers", *IEEE Journal of Solid-State Circuits*, Vol. SC-9, 1974, pp. 353-364.

8. Texas Instruments, *Operational Amplifier Macromodels Data Manual*, 1990.

9. Biagi, H., & Stiff M.R., *Operational Amplifier and Instrumentation Macromodels*, Application Bulletin, Burr-Brown Corporation, 1990.

10. Buxton, J., "Improve Noise Analysis With Op-Amp Macromodel", *Electronic Design*, April 1992, pp. 73-81.

11. Faehnrich, R., "Op-Amp Model Includes 1/f Noise", *EDN Magazine*, April 23, 1992, pp. 172-173.

12. Comlinear Corporation, "A New Approach to Op-Amp Design", Application Note 30-1, March 1985.

13. Al-Hashimi, B.M., "Rewriting the Rules with Current Mode Amplifiers", *Electronics World + Wireless World*, October 1993, pp. 843-846.

14. Elantec Inc, *Data Book 1992*, Milpitas, CA.

15. Hageman, S., "Improved Simulation Accuracy When Using Passive Components", The Design Center Source, MicroSim Corporation, April 1994, pp. 8-12.

chapter seven

Sensitivity and Monte Carlo analyses

Practical circuit performances vary from the ideal because component parameters vary typically due to tolerance, temperature, or aging. These variations can be classified according to whether they can be accurately predicted (deterministic) or whether statistical methods must be applied. For example, the base-emitter voltage variation with temperature of a bipolar transistor could be taken as a deterministic change (at approximately –2mV/°C), but the actual base-emitter voltage for a given transistor at a constant temperature could only be known with statistical limits due to the production variables. This chapter introduces the statistical analyses available in PSpice. There are two analyses: the first is sensitivity and worst-case, and the second is Monte Carlo analysis. Worked examples of amplifier and filter circuits will be given to illustrate the use of both analyses.

Before explaining how PSpice performs the statistical analyses, it is necessary to introduce some basic definitions and show, for example, how they are used to calculate the sensitivity of a circuit.

7.1 Sensitivity definitions

As an illustration, consider the noninverting amplifier circuit shown in Figure 7.1.

The voltage transfer function (F) of the amplifier is

$$F = \frac{V_o}{V_i} = \frac{R1 + R2}{R1} \tag{7.1}$$

Mathematically, the sensitivity of F with respect to x is

$$S_x^F = \frac{x}{F} \frac{\partial F}{\partial x} \tag{7.2}$$

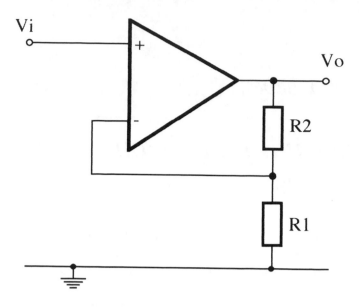

Figure 7.1 Amplifier circuit.

where F is the transfer function of the circuit, while x can be a circuit component parameter. The sensitivity of F to the component R1 is

$$S_{R1}^F = \left(\frac{R1}{F}\right)\frac{\partial F}{\partial R1} \tag{7.3}$$

Differentiating Equation (7.1) with respect to R1 gives

$$\frac{\partial F}{\partial R1} = \frac{-R2}{(R1)^2} \tag{7.4}$$

Substituting Equations (7.1) and (7.4) into Equation (7.3) gives

$$S_{R1}^F = \frac{-R2}{R1+R2} \tag{7.5}$$

Similarly, the sensitivity of F to the component $R2$ is

$$S_{R2}^F = \frac{R2}{R1+R2} \tag{7.6}$$

If R1=R2=R, then $S_{R1}^F = -0.5$, and $S_{R2}^F = 0.5$. This means that a 1% increase in R1 results in the amplifier gain decreasing by 0.5%, while a 1% increase in R2 results in the amplifier gain increasing by 0.5%.

In this example, the result can be verified by substitution of actual values and tolerances. Assume R1=R2=1kΩ, the gain of the amplifier is

(R1+R2)/R1=2000/1000=2

If R1 varies by 1% and becomes 1010Ω, the gain is then

(1010+1000)/1010=1.99

This is a reduction in gain of 0.5%, as predicted.

For complex circuits, manual calculation of sensitivity and worst-case expressions is extremely tedious. PSpice allows sensitivity and worst-case analyses by assigning the appropriate tolerance to each component.

7.2 Component tolerances

Component tolerances can behave independent of each other, or there may be a relationship between similar components. PSpice allows the designer to choose whether a component behaves independently by use of the DEV parameter, or the component tolerance track by the use of the LOT parameter. PSpice uses the .MODEL statement to assign a component with tolerances. For example,

R1 NODE1 NODE2 RMOD 1K
R2 NODE2 NODE3 RMOD 2K
.MODEL RMOD RES (R=1 DEV=5%)

Here both R1 and R2 use the model name RMOD, with a nominal value for R (the scaling factor) of 1. Incidentally, the model name RMOD has been chosen arbitrarily and has no meaning itself. During the various runs of a statistical analysis, the values of R1 and R2 will vary by 5% (at most). With a DEV tolerance, this variation is independent. So, for a simulation run, R1 could be any value between 950 and 1050Ω while R2 could be any value between 1900 and 2100Ω.

The LOT tolerance means that the components that use the same .MODEL statement will vary together. For example, if the .MODEL statement in the above example is changed to

.MODEL RMOD RES (R=1 LOT=5%)

Now R1 and R2 will always track each other, which means that the values of R1 and R2 will increase or decrease by the same percentage. The two types of tolerances (DEV and LOT) can also combine; for example, modifying the previous .MODEL statement to

.MODEL RMOD RES (R=1 LOT=1% DEV=2%)

means that the two tolerances add. For each run, R is first assigned a LOT variation up to 1% and each resistor is assigned a DEV variation up to 2%, in addition to the LOT variation. Therefore, each resistor could be as much as 3% from nominal values. Combined LOT and DEV tolerances are useful for situations where component variations are not completely correlated, but are not completely independent either. In this case the DEV=2% could deal with the noncorrelated initial tolerance, while the LOT=1% could represent an expected change due to temperature drift.

7.3   Sensitivity and worst-case analyses

The prime purpose of sensitivity and worst-case analyses is to identify the most critical components of a circuit. During sensitivity analysis, components are varied *one* at a time, and an analysis is obtained for each variation. This means that sensitivity analysis allows PSpice to find the sensitivity of the circuit output to each component. After each component has been varied, a worst-case analysis is done with *all* components varied to give the worst-case output.

The amount by which each component is varied during the sensitivity analyses is predetermined by the designer and set by the parameter RELTOL (relative tolerance) according to the equation

New value=nominal value*(1+RELTOL)

PSpice defaults RELTOL to 0.001 or 0.1%, but this can be overridden using the .OPTIONS command. For example, including the statement

.OPTIONS RELTOL=0.01

in the circuit input file causes PSpice to assign each component a tolerance of 1% during the sensitivity analyses. Having determined the sensitivity of the circuit output to each component, component tolerance values specified by the user (DEV or LOT) in the .MODEL statement will be used to obtain the worst-case result. This result is obtained when each component is varied up or down by its *full* tolerance value.

PSpice performs both sensitivity and worst-case analyses using the following statement:

.WCASE *<analysis> <output variable> <function> [option]*

The full description of this statement is given in Box 7.1.

Box 7.1 PSpice Sensitivity and Worst-Case Analyses Commands

PSpice performs sensitivity and worst-case analyses using the command .WCASE, the general form of which is

.WCASE *<analysis> <output variable> <function>* [*option*]

The parameter *<analysis>* defines the type of analysis required. This could be an AC, DC, or transient analysis; *<output variable>* defines the type of output variables; *<function>* specifies the operation to be performed on the values of the *<output variable>* to reduce these to a single number. This number is then used as the basis for the comparison between the nominal and subsequent runs. This is used to compress the results of sensitivity and worst-case analyses. There are five *<functions>* available:

1. YMAX finds the greatest difference in the Y direction in each waveform from the nominal run.
2. MAX finds the maximum value of each waveform.
3. MIN finds the minimum value of each waveform.
4. RISE_EDGE (*value*) finds the first occurrence of the waveform crossing above the threshold *<value>*. The waveform must have one or more points below *<value>* followed by one above; the output value listed will be where the waveform increases above *<value>*.
5. FALL_EDGE (*value*) finds the first occurrence of the waveform crossing below the threshold *<value>*. The waveform must have one or more points at or above *<value>* followed by one below; the output value listed will be where the waveform decreases below *<value>*.

Only *one* function must be selected when sensitivity and worst-case analyses are performed. This function is usually called the *collating function*.

There are eight [options] available, of which only three will be given here. These particular options are often considered to be the most useful. For information on the other options, see the PSpice User's Guide Manual[1]. The three options are

1. LIST will print the updated model parameters used in the sensitivity analysis.
2. OUTPUT ALL requests output from the sensitivity runs, after the nominal (first) run. If OUTPUT ALL is omitted, then only the nominal and worst-case runs produce output.
3. DEVICES — By default all components are included in the sensitivity and worst-case analyses. The components considered may be limited by listing the components after the keyword DEVICES. For example, to perform analysis only on resistors and capacitors in a circuit, enter

DEVICES RC.

PSpice often generates a large amount of information when performing sensitivity and worst-case analyses. To represent the results in the most useful format, PSpice provides a number of summary reports in the circuit output file (i.e., .OUT file). These reports are as follows:

1. **Updated Model Parameter Summary Report** — This report shows the actual component values used in each of the sensitivity analyses. The component values depend on the value of the parameter RELTOL. This report is generated if the LIST option is included on the .WCASE command line.

2. **Sensitivity Summary Report** — This report gives a summary of all the outputs from the sensitivity analyses using a collating function (see Box 7.1 for details of collating functions). Usually the output of each analysis is presented by a single number. Depending on the chosen collating function, this number could describe, for example, the output maximum deviation from the nominal. This report is important since it provides the sensitivity of the circuit output to each component. PSpice generates this report as the default report.

3. **Worst-Case Summary Report** — In this summary, a list of all the component values used in the worst-case analysis is given, with each component varied up or down by its full tolerance. Also included in this report, depending on the selected collating function, is the worst-case analysis output. PSpice generates this summary report by default.

7.4   Simulation examples

Two examples are considered, the first of which is based on a simple amplifier circuit. This example is chosen so that a direct comparison between the simulated and calculated results obtained in Section 7.1 can be made. The second example is more complex and is based on a bandpass filter. This is intended to show how PSpice can help the designer identify the critical components on the filter response. Also a direct comparison between the simulated results and the commonly known filter sensitivity expressions can be made.

Example 7.4.1 Amplifier Circuit
The amplifier circuit is shown in Figure 7.1, and is repeated here for convenience with nodes in Figure 7.2. It is assumed that the amplifier has a gain of 2 and is obtained by setting R1=R2=1kΩ. The PSpice input file of the amplifier circuit is given in Listing 7.1.

During sensitivity analyses, both resistors will have the same tolerance, which is set by the RELTOL parameter value, and in this example

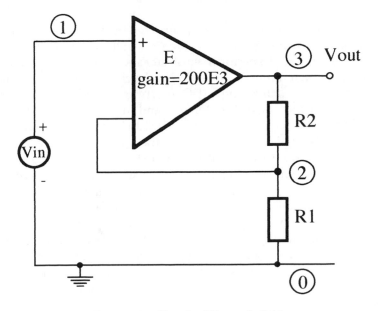

Figure 7.2 Circuit of Example 7.4.1.

it is 1%. The resistors are modelled using the .MODEL statement, and it is assumed that they have a 1% tolerance and vary independent of one another (i.e., DEV tolerance). This tolerance value will be used in the worst-case analysis.

Listing 7.1 PSpice Input File of Example 7.4.1

```
Amplifier Circuit (Figure 7.2); title line
*
.OPTIONS RELTOL=0.01       ; 1% resistors (sensitivity analyses)
*
Vin 1 0                    ; input source
*
.DC Vin 0V 1V 1V           ; DC sweep analysis
*
R1 2 0 RMOD 1K             ; modelled resistor
*
R2 2 3 RMOD 1K
*
E 3 0 1 2 200E3            ; ideal amplifier model
RL 1 2 1E9                 ; eliminate floating nodes
*
.MODEL RMOD RES (R=1 DEV=1%); 1% resistors (worst case)
*
.WCASE DC V(3) YMAX        ; sensitivity & worst-case command
*
.END                      ; end of input file
```

To simplify the analysis, the op-amp has been assumed to be ideal and described using a voltage-controlled voltage source (i.e., the E component model). To perform sensitivity and worst-case analyses, the .WCASE command must be included in the circuit input file as shown in Listing 7.1. An analysis type (DC, AC, or transient) should be specified as part of the .WCASE statement, and in this example, it is a DC sweep that starts at 0V and finishes at 1V with step of 1V. PSpice always performs a nominal (components set to their ideal values, i.e., no tolerance) analysis prior to sensitivity and worst-case analyses. To compare the results between the nominal and the sensitivity analyses, a collating function must be specified on the .WCASE command line. In this example the function YMAX has been chosen since it provides the maximum deviation of the amplifier output, V(3), in the Y direction from the nominal output. Box 7.2 shows the PSpice sensitivity report summary of the amplifier circuit. This report is part of the circuit output file. This shows in the sensitivity run or analysis of R2; the amplifier output voltage (or gain) has a deviation of 0.01. This deviation is "higher" than the nominal value and is occurring at Vin=1 (input signal value). This means that there is a 0.5% increase (change) in the amplifier gain for every 1% increase (change) in R2. In the sensitivity run of R1, the amplifier gain has a deviation of 0.0099. This deviation is "lower" than the nominal value and is occurring at Vin=1. This means that there is a 0.495% decrease in the amplifier gain for every 1% increase in R1. This shows that within numerical calculation limits, there is a good agreement between PSpice and the analytical calculation of the amplifier sensitivity given in Section 7.1. The worst-case summary report of the amplifier circuit is given Box 7.3. This report, which is part of the circuit .OUT file, shows that at worst case the amplifier gain would have a maximum deviation of 0.0202. This deviation is "higher" than the nominal value and is occurring at Vin=1. As mentioned earlier, the

Box 7.2   PSpice Sensitivity Summary Report of Example 7.4.1

```
****  SORTED DEVIATIONS OF V(3)           TEMPERATURE= 27.0 DEG C


                     SENSITIVITY SUMMARY
********************************************************************

RUN                     MAX DEVIATION FROM NOMINAL

R2 RMOD R.01  (1.01 sigma) higher at Vin = 1

        (.5% change per 1% change in Model parameter)


R1 RMOD R 9.9009E-03 (1.00 sigma) lower at Vin = 1

        (.495% change per 1% change in Model parameter)
```

tolerance value used in the sensitivity analyses is set by the parameter RELTOL, while the tolerance values used in the worst-case analysis are set by the DEV or LOT parameters specified in the .MODEL statement. During worst-case analysis, R1 and R2 will be varied up or down by their full specified tolerance value. To determine in which direction each component tolerance should be set, the output value from each sensitivity analysis is evaluated against that from the nominal analysis. Then, depending upon the specified collating function, PSpice decides whether to set the component tolerance to maximum or minimum. In Box 7.3, R1 was set to its minimum allowed value, while R2 was set to its maximum allowed value since these conditions provide the maximum deviation from the nominal output as required by the collating function.

Example 7.4.2 Bandpass Active Filter

Consider a requirement for a bandpass filter with a centre frequency (F_o) of 10kHz, 3dB bandwidth of 2kHz, and a gain (K) of 1. The required filter is shown in Figure 7.3, and it is based on the popular multiple loop feedback (MLF) configuration[2].

The PSpice input file of the circuit is given in Listing 7.2. Note how the capacitor tolerance is specified using the .MODEL CMOD CAP command. It has been assumed that 5% tolerance resistors and 10% tolerance capacitors will be used in realisation of the filter. Also the component tolerances have been assumed to vary independently. Filters are often characterised by various performance parameters, including, 3dB bandwidth, centre frequency, passband ripple, stopband attenuation, and gain in the case of active filters. All these parameters are

Box 7.3   PSpice Worst-Case Summary of Example 7.4.1

```
****WORST-CASE ANALYSIS TEMPERATURE=27.000 DEG C

                  WORST-CASE ALL DEVICES

*************************************************************
****UPDATED MODEL PARAMETERS TEMPERATURE=27.000 DEG C

                  WORST-CASE ALL DEVICES

*************************************************************
DEVICE      MODEL       PARAMETER       NEW VALUE
R1          RMOD        R               .99   (Decreased)
R2          RMOD        R               1.01  (Increased)

                  WORST-CASE SUMMARY
*************************************************************
RUN                     MAX DEVIATION FROM NOMINAL

ALL DEVICES                 .0202 higher at Vin=1
```

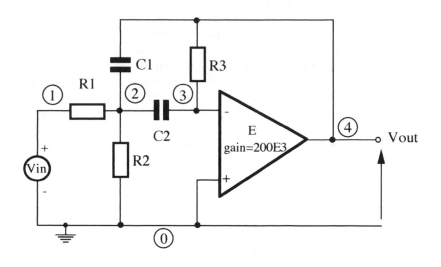

Figure 7.3   Circuit of Example 7.4.2.

therefore important, depending on the application. This makes the comparison of sensitivity analyses of filters a difficult task. To simplify the comparison, in this example, the centre frequency (F_o) and the gain (K) of the filter will be assumed to be important, and therefore will be used as the basis for comparison between the various sensitivity analyses. The function MAX will be chosen as the collating function since it gives the maximum value of the filter output (i.e., gain) for each sensitivity analysis. To determine the effect of each filter component on the circuit response, the sensitivity summary report is required. This report, which is part of the circuit .OUT file, is given in Box 7.4. This shows, for example, in the sensitivity analysis of R3, that the filter has a maximum gain of 1.01 at a frequency of 9.95kHz. This represents the filter centre frequency since the centre frequency of an active filter is usually defined as the frequency point at which the gain of the filter is maximum. This means that there is a 1% increase (change) in the filter gain for every 1% increase (change) in R3. Similarly, in the sensitivity analysis of C1, the filter has a maximum gain of 0.995 and a centre frequency of 9.95kHz. This means that there is a 0.4975% decrease in the filter gain for every 1% increase in C1. Note that during the nominal analysis, PSpice shows that the filter has a maximum gain of 1 and centre frequency of 9.999kHz rather than the theoretical 10kHz. This error in frequency is due to a combination of the rounding of the calculated filter component values and the numerical accuracy of PSpice.

Listing 7.2 Input File of Example 7.4.2

```
10kHz Bandpass Filter (Figure 7.3) ; title line
*
.OPTIONS RELTOL=0.01            ; 1% components (sensitivity run)
*
Vin 1 0 AC 1                    ; AC analysis
*
.AC LIN 1000 9.5kHz 10.5kHz     ; Frequency range
*
R1 1 2 RMOD 0.796K              ; modelled resistor
*
R2 2 0 RMOD 16.24
*
R3 3 4 RMOD 1.592K
*
C1 2 4 CMOD 0.1uF               ; modelled capacitor
*
C2 2 3 CMOD 0.1uF
*
E 4 0 0 3 200E3                 ; Ideal op-amp description
*
.MODEL RMOD RES (R=1 DEV=5%)    ; 5% resistors (worst case)
*
.MODEL CMOD CAP (C=1 DEV=10%)   ; 10% capacitors (worst case)
*
.WCASE AC V(4) MAX OUTPUT ALL   ; sensitivity & worst case
*
.PROBE V(4)                     ; graphic filter output
*
.END                           ; end of input file
```

The sensitivity analysis of the MLF bandpass filter has been studied in detail over the years, and it is well covered in many filter design text books[2]. To simplify the comparison between PSpice MLF filter sensitivity results and that given in Reference 2, the information contained in Box 7.4 has been tabulated as shown in Table 7.1, where S_x^K, for example, represents the sensitivity of the filter gain (K) to a particular circuit component, x. The S_x^{Fo} values were obtained by calculating the percentage difference between the centre frequency of each component sensitivity analysis and the filter nominal centre frequency of 9.999kHz. For example, in the sensitivity analysis of C1, Box 7.4 shows that the filter has a centre frequency of 9.95kHz. Therefore, the percentage difference is (9.95–9.999)*100/9.999=–0.49%. The sensitivity results given in Table 7.1 are in good agreement with the classical sensitivity expressions of the MLF bandpass filter given in Reference 2. Note that the sensitivity results given in Table 7.1 have been obtained assuming all the filter components have a tolerance of 1% (the RELTOL parameter value).

Box 7.4 PSpice Sensitivity Summary Report of Example 7.4.2

```
** SORTED DEVIATIONS OF V(4)    TEMPERATURE=27.0 DEG C

                      SENSITIVITY SUMMARY
* * * * * * * * * * * * * * * * * * * * * * * * * * * * * * * * * * * * * * * * * * * * * * * * * * * * * * * *

RUN                           MAXIMUM VALUE

R3 RMOD R                      1.01    at F =    (9.9500E+03)

    (1% change per 1% change in Model Parameter)


C2 CMOD C                      1.005   at F =     9.9500E+03
         (.4976% change per 1% change in Model Parameter)


NOMINAL                        1       at F =     9.9990E+03


R2 RMOD R                      1       at F =     9.9500E+03
         (-41.7240E-06% change per 1% change in Model Parameter)


C1 CMOD C                      .995    at F =     9.9500E+03
         (-.4975% change per 1% change in Model Parameter)


R1 RMOD R                      .9901   at F =     9.9990E+03
         (-.99% change per 1% change in Model Parameter)
```

Thus far, only the filter sensitivity results have been discussed. To obtain the worst-case output, the worst-case summary report is needed. This report, which is part of the .OUT file, is given in Box 7.5. This shows that at worst the filter will have a maximum gain of 1.2158 and a centre frequency of 10.062kHz. Note that the worst-case result is obtained when the resistor and capacitor values have been set to their maximum or minimum tolerances values (5% resistors, 10% capacitors), depending on the filter gain during each sensitivity analysis. For example, in the sensitivity analysis of C1, Box 7.4 shows that the filter has a maximum gain of 0.995, which is less than the nominal gain of 1.

Table 7.1 PSpice Sensitivity Results of the MLF Bandpass Filter with 1% Component Tolerances

Component x	S_x^K	$S_x^{F_o}$
C1	−0.4975	−0.490
C2	+0.4976	−0.490
R1	−0.99	0
R2	−41.72E−06	−0.490
R3	+1	−0.490

Box 7.5 PSpice Worst Case Summary Report of Example 7.4.2

```
****WORST-CASE ANALYSIS          TEMPERATURE =  27.000 DEG C

                    WORST-CASE ALL DEVICES

*******************************************************************
****UPDATED MODEL PARAMETERS TEMPERATURE =  27.000 DEG C

                    WORST-CASE ALL DEVICES

*******************************************************************

DEVICE      MODEL      PARAMETER      NEW VALUE
C1          CMOD       C                  .9  (Decreased)
C2          CMOD       C                 1.1  (Increased)
R1          RMOD       R                  .95 (Decreased)
R2          RMOD       R                  .95 (Decreased)
R3          RMOD       R                 1.05 (Increased)


*******************************************************************

****SORTED DEVIATIONS OF V(4)   TEMPERATURE =  27.000 DEG C

                    WORST-CASE SUMMARY

*******************************************************************

RUN                      MAXIMUM VALUE

ALL DEVICES              1.2158 at F =   10.0620E+03
NOMINAL                  1      at F =    9.9990E+03
```

This means that during the worst-case analysis the capacitor C1 will be set to its minimum tolerance value (i.e., 0.9 of its nominal value) as shown in Box 7.5. Similarly, in the sensitivity analysis of C2, the filter has a maximum gain of 1.005, which is greater that the nominal gain. Therefore, C2 will set to its maximum tolerance value as shown in Box 7.5.

The number of sensitivity analyses performed by PSpice depends on the number of components in a circuit. For example, in Listing 6.2 PSpice will perform five sensitivity analyses (one for each filter component), as well as the default nominal and worst-case analyses, making seven analyses in total. To plot the results of all these analyses using *Probe*, the option OUTPUT ALL must be specified as part of the .WCASE command statement as shown in Listing 6.2. Figure 7.4 shows the frequency response of the bandpass filter for the nominal and each of the filter component sensitivity analyses, where V(4)@1, for example, represents the nominal analysis. The response is shown over the frequency range of 9.5 to 10.5kHz to illustrate clearly the effect of each component on the filter response. Figure 7.5 shows the nominal and the

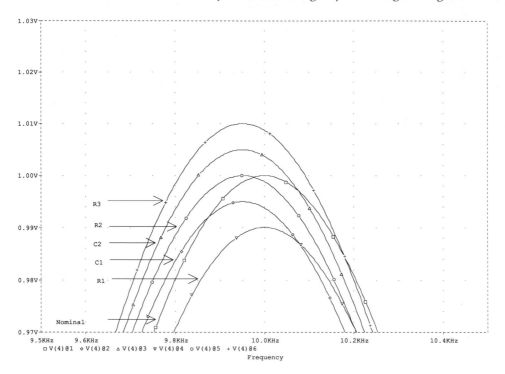

Figure 7.4   Frequency response of circuit (Figure 7.3) for various sensitivity runs.

worst-case response of the 10kHz bandpass filter, where V(4)@7 represents the worst-case analysis.

It should now be clear that the selection of the collating function is very important. PSpice is capable of performing sensitivity and worst-case analyses on a large number of parameters. For example, in the bandpass filter just discussed, the 3dB bandwidth may be the important factor. To perform sensitivity analyses on this design parameter, the collating functions RISE_EDGE and FALL_EDGE (value) must be used. The (value) parameter of the collating function should be 0.707, which is equivalent to the 3dB point. It should be noted that PSpice allows only one collating function to be selected during the sensitivity and worst-case analyses. This means that to compare the filter in terms of the –3dB bandwidth, two separate analyses must be performed. One analysis, with the collating function, is RISE_EDGE (0.707), which is the upper –3dB frequency point; another, with the collating function, is FALL_EDGE (0.707), which is the lower –3dB frequency point. The results from the two analyses are then subtracted to yield the –3dB filter bandwidth.

The two examples given show that PSpice is quite capable of performing sensitivity and worst-case analyses without the need to derive

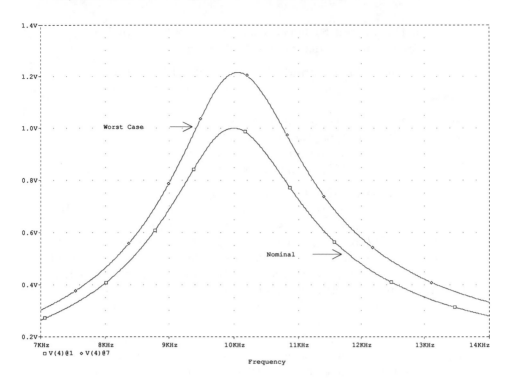

Figure 7.5 Frequency response of circuit (Figure 7.3) for nominal and worst-case runs.

time-consuming and error-prone sensitivity expressions. The key to success when using PSpice in this mode is defining the collating functions such that the results provided can be properly interpreted. For really complex circuits or circuits with a large number of interacting and interdependent parameters PSpice will provide the results, but the interpretation is still firmly with the designer.

7.5 Monte Carlo analysis

This is the second type of statistical analysis available in PSpice. This type of analysis is often used before designs are put into volume production, where the manufacturing process is simulated with factors like component tolerances. A number of test circuits, typically 100 to 300, are manufactured on the computer, with component values picked at random from their respective statistical distributions. During a Monte Carlo analysis run *all* components are varied, unlike the procedure during sensitivity analysis, when components are varied one at a time during an analysis. Component tolerances are assigned using the DEV and LOT parameters, the same as in the sensitivity and worst-case analyses. The amount by which each component is varied during a Monte Carlo analysis is determined by PSpice:

New Value = Nominal Value. (1+ Rand.Tolerance)

where:

New value = updated component value.
Nominal value = original component value.
Rand = a number chosen from the random number generator.
Tolerance = DEV or LOT tolerance specified in the .MODEL statement.

The component tolerances are varied randomly according to a specified statistical distribution. Pspice has three tolerance distributions: uniform, Gaussian, and user defined. The simplest distribution is the uniform (Figure 7.6a), in which components are equally likely to take on any value with the tolerance band. PSpice defaults component tolerances to the uniform distribution. The most common distribution in manufacturing is often considered to be the Gaussian, or normal, distribution (Fig. 7.6b). A Gaussian distribution is described by two parameters: the mean, usually represented by $\bar{x}$, and the standard deviation, σ. The standard deviation of a component is a measure of the departure of the component from its mean value. The larger the standard deviation, the greater the departure from the mean value. There are some important points to note about the Gaussian curve. The first is that the probability of the component's lying between two given values is found from the area under the curve between those values. For instance, the area under the curve between $\bar{x}$ and ($\bar{x}+\sigma$) is 0.34, so the probability of the component's being between the mean and one standard deviation above the mean is 0.34. The second point to note is that the Gaussian curve is symmetrical about the mean value. The third point is that the Gaussian curve is nearly zero beyond values of ($\bar{x}\pm3\sigma$). In other words, the probability of finding a value more than three

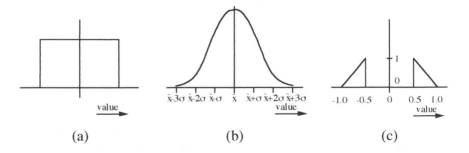

(a)   (b)   (c)

Figure 7.6 Component tolerance distributions available in PSpice: (a) uniform; (b) Gaussian; (c) user defined.

standard deviations away from the mean is very small. PSpice limits values of Gaussian distributions to ±4σ.

Example 7.5.1

To specify a Gaussian distribution for a ±1% resistor, the following PSpice statements are required:

```
.OPTIONS DISTRIBUTION=GAUSS; Gaussian Distribution
.MODEL RMOD RES (R=1 DEV/GAUSS=0.25%); ±1% resistors
```

where the resistor with the model name RMOD has a Gaussian distribution and σ=0.25%. The resistor tolerance value is determined by multiplying σ by 4 (i.e., ±1% tolerance). Often in practice, component tolerances do not follow uniform or Gaussian distributions, and PSpice supports a user-defined distribution, where the distribution is specified as an array of (*<deviation>*,*<probability>*) pairs, with up to 100 value pairs allowed.

Example 7.5.2

To specify a five-part distribution[3] for ±20% capacitors (Figure 7.6c), the following statements are needed:

```
.OPTIONS DISTRIBUTION=USERDEF; user defined distribution
.DISTRIBUTION USERDEF  (-1,0)  (-0.5,1)
+                      (-0.5,0)  (0.5,0)
+                      (0.5,1)  (1,0)  ; distribution graph
.MODEL CMOD CAP (C=1 DEV/USERDEF=20%)  ; ±20% capacitors
```

where a capacitor with a model name CMOD has been assumed.

Monte Carlo analysis is performed using the command .MC. This command has many similarities with the sensitivity and worst-case analyses command .WCASE (Box 7.1). The general form of the .MC command is

.MC *<runs value> <analysis> <output variable> <function> [option]*

The main difference between the two analyses is that in Monte Carlo analysis, the user must specify the number of runs required, as shown above in the .MC command. Unlike the sensitivity and worst-case analyses, when the user does not need to specify the number of runs because it is determined by the number of model parameters to vary and the number of components that reference these models. For instance, in Example 7.4.1, PSpice will do a total of four analyses — one nominal, two sensitivity (1 for R1 and 1 for R2, varying R for each), and one worst case. The maximum number of runs allowed in PSpice is 2000, and if the output is to be viewed with the command .PROBE (i.e., graphic output), the limit is 100 runs. There is a trade-off in choosing the number of Monte Carlo runs. More runs provide better statistics, but take proportionally more computer time. It should be noted that as the number of runs increases, the more closely PSpice approximates the actual worst-case limits of the circuit.

Example 7.5.3
To perform a Monte Carlo analysis on the bandpass filter considered in Example 7.4.2, the .WCASE command in Listing 7.2 needs to be replaced by the command:

.MC 100 AC V(4) MAX OUTPUT ALL

This command causes PSpice to do 100 runs, one nominal and 99 Monte Carlo runs. During the various runs, the filter components (resistors and capacitors) will vary according to the default uniform tolerance distribution. If other distribution is required, then it must be specified in the circuit input file, as discussed in Examples 7.5.1 and 7.5.2.

Here, the various Monte Carlo analyses will be compared using the collating function MAX. Note that the OUTPUT ALL option specified on the .MC command will cause 100 graphs of the filter frequency response to be generated. This output option is the only one available in sensitivity and worst-case analyses. However, it is possible to change this output option in Monte Carlo analysis; PSpice offers three extra options:

OUTPUT FIRST (value)
OUTPUT EVERY (value)
OUTPUT RUNS (value)

The first option produces output for only the first (value) run, the second for every (value) run, and the last for whichever runs are listed. PSpice also has the facility of producing histograms of data obtained from Monte Carlo analysis (see page 232 of Reference 1 for more information on creating histograms).

Finally Monte Carlo analysis has the same collating functions as the sensitivity and worst-case analyses, and it generates the same summary output reports (these reports were discussed in Section 7.3).

As for sensitivity and worst-case analysis, PSpice provides excellent results from its Monte Carlo analysis. The designer still has to define the distribution functions for the components, and this requires considerable understanding of correct manufacturing and selection procedures. The designer is responsible for choosing the collating functions and also interpreting the results.

7.6 Temperature analysis

Very often, it is necessary to check a circuit performance at different temperatures to verify that certain circuit design parameters are met. PSpice allows the user to analyse circuit performance at various temperatures using the command

.TEMP <temperature value>

where the temperature value is in degrees Centigrade. PSpice defaults to assumed room temperature of 27°C. The parameter <*temperature value*> can be a single value or a list of values. For example, including the statement

.TEMP 20 80 –20

in the circuit input file will cause PSpice to do three analyses, one at 20°C, one at 80°C, and one at –20°C.

It is also possible to sweep the temperature parameter over a range of values using the .STEP command (see Section 5.3 of Chapter 5). For example the statement

.STEP TEMP 20 90 10

will direct PSpice to perform eight temperature runs in steps of 10°C starting at 20°C and finishing at 90°C.

In order to simulate circuits at various temperatures, the components of the circuit must have variation in their values. Passive components such as resistors, inductors, and capacitors have a PSpice built-in temperature-dependence model, which for the resistor is

Resistance=<Value>.R.$(1+TC1.(T-T_0)+TC2.(T-T_0)^2)$

where T_0 is the default PSpice temperature of 27°C, T is the required temperature, TC1 is the linear temperature coefficient per degree Centigrade, and TC2 is the quadratic temperature coefficient per degree Celsius squared. The temperature model for the capacitors and inductors follows the same equation. The user must specify the values of TC1 and/or TC2 in the .MODEL statement if passive component variation with temperature is required.

Example 7.6.1
To model a 10µF capacitor with a linear temperature coefficient of –500ppm/°C, the following statements are needed:

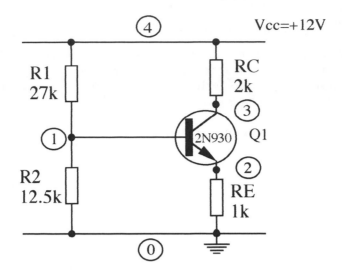

Figure 7.7   Circuit of Example 7.6.2.

C 2 0 CMOD 10uF
.MODEL CMOD CAP (C=1 TC1=–0.00050)

Active components such as transistors have temperature-dependency factors built into their models, and the user does not need to provide such factors. Others, such as op-amp models, may or may not have temperature-dependency factors, depending on the quality of the models supplied by the op-amp manufactures (see Chapter 6, Section 6.5).

Example 7.6.2
Find the DC biasing points of the common emitter amplifier (Figure 7.7) when operated at 27, 90, and –30°C. It is assumed that the transistor type is the Q2N930 and that all the resistors are of the metal-film type with a temperature coefficient of 200ppm/°C.

The input file of the transistor circuit is given in Listing 7.3. A model of the transistor (Q2N930) is obtained from the PSpice "BIPOLAR.LIB" library (See Table 6.6 of Chapter 6). Note that the various temperature analyses are performed using the .TEMP command and that the temperature coefficient is included in the .MODEL statement. Table 7.2 shows the simulated amplifier DC biasing points for the various temperatures. Table 7.3 gives the resistor values used in the various analyses. Both tables were derived from information given in the PSpice circuit output file. PSpice also shows in the output file how the various transistor parameters vary with temperature such as the current gain (β). Table 7.4 shows how β varies with temperature in the transistor model.

Listing 7.3   PSpice Input File of Example 7.6.2

```
Common Emitter Amplifier Circuit (Figure 7.7)   ; Title line
*
.TEMP 27 90 -30              ; temperature range
*
Vcc 4 0 +12V                 ; positive power supply
*
R1 1 4 RMOD 27K              ; modelled resistor
*
R2 1 0 RMOD 12.5K
*
RE 2 0 RMOD 1K
*
RC 3 4 RMOD 2K
*
.MODEL RMOD RES (R=1 TC1=0.00020); resistor temp model
*
Q1 3 1 2 Q2N930             ; transistor connections & type
.LIB C:\BIPOLAR.LIB         ; bipolar transistor model library
*
.END
```

7.7   Chapter summary

There are two statistical analyses methods available in PSpice: the first is sensitivity and worst-case analysis, and the second is Monte Carlo analysis. Component tolerances are assigned using the DEV and LOT parameters. With sensitivity analysis, components are varied one at a time. The amount by which each component is varied is predetermined by the user and set by the RELTOL parameter. PSpice defaults RELTOL to 0.001 or 0.1%. After each component has been varied, one final run is made with all the components varied to give the worst-case output.

In Monte Carlo analysis, all components are varied randomly according to a selected statistical tolerance distribution. PSpice has three distributions: uniform, Gaussian, and user defined, with the default being the uniform distribution.

Table 7.2   Amplifier DC Biasing Points at
Different Temperatures

Temperature (°C)	V(1)	V(2)	V(3)
27	3.704	3.0036	6.0167
90	3.724	3.125	5.767
−30	3.675	2.884	6.261

Table 7.3   Resistor Values at Different Temperatures

Temperature (°C)	R1(kΩ)	R2(kΩ)	RE(kΩ)	RC(kΩ)
27	27	12.5	1	2
90	27.34	12.66	1.013	2.025
−30	26.69	12.36	0.9886	1.977

Table 7.4   Transistor β Values at Different Temperatures

Temperature (°C)	β
27	578
90	770
−30	412

PSpice allows the user to analyse circuits at various temperatures using a combination of commands .TEMP and .STEP.

PSpice statistical analyses provide very powerful tools for predicting actual circuit performance, but the final usefulness of this option of analyses is dependent upon good input data (in terms of component tolerances, etc.) and good selection of collating functions to allow the PSpice output to be correctly interpreted.

References

1. MicroSim Corporation, *Circuit Analysis — User's Guide Manual* (The Design Center), Version 5.3, Irvine, CA, January 1993.
2. Stephenson, F. W., *RC Active Filter Design Handbook*, John Wiley & Sons, New York, 1986, Chapter 11 (p. 309).
3. Ellis, G., "Use Spice to Analyze Component Variations in Circuit Design", *EDN Magazine*, April 29, 1993, pp. 109-114.

chapter eight

Analogue behavioural modelling "block-diagram" simulation approach

Analogue behavioural modelling (ABM) allows the designer to model analogue circuit functions using mathematical equations, tables, and transfer functions. The designer can then simulate systems as a combination of "block diagrams", each of which performs a specific function. The aim of this chapter is to show how behavioural models of analogue circuits are developed. An AM modulator/demodulator system will be used as an illustration of some of the models available.

This chapter also shows how ABM can be used to model practical components. As an example, the modelling of the gain-frequency characteristics of op-amps will be considered.

8.1 Introduction

Circuit simulators allow the prediction of system performance. To carry out the simulation, a detailed component design of the system must be available. The system is then described to the simulator using a basic set of components, including resistors, transistors, and various voltage and current sources. The connections of these components are then expressed in terms of nodes. This type of simulation is often called the structure or primitive-level simulation. At the circuit level, clearly primitive-level simulation is required, and this kind of simulation has been the type used up to now. When a designer is operating at the system level, producing component-level simulation may be too detailed. At this level, a block-diagram simulation approach may be more appropriate, where specific functions may be described by their mathematical behaviour. For example, let us assume that part of the system under simulation is a second-order lowpass filter. In order to optimise the system performance, the effect of changing the filter type (Butterworth, Chebyshev, etc.) on the system response needs to be examined. Until recently, to do this a number of different types of filters would have had to be designed and simulated at the component level, which could prove to be a time-consuming task. A more effective way of dealing with this problem is to simulate the filter as a block-diagram, where the input/output relationship can be expressed in terms of an equation. This means that the filter component design is unnecessary at this stage, since the filter function has been modelled using a mathematical expression. At a later stage this model can be replaced by actual circuitry. This type of simulation is called behavioural-level simulation, where a functional block can be described by its behaviour without worrying about its physical structure. The description can be an equation or a data table.

This type of simulation enables the user to check and optimise a *system design* without the need to perform *circuit design*. It also allows complex systems to be simulated quickly and more efficiently when compared with the primitive simulation level.

8.2   Implementation

PSpice performs ABM by extending the capabilities of the two controlled sources (VCVS) and (VCCS). The VCVS is a voltage-controlled voltage source, while the VCCS is a voltage-controlled current source. These sources, which were discussed in Chapter 4 (Sections 4.1.1 and 4.1.2), are identified by PSpice as components, starting with the letters E and G, respectively. Therefore, when modelling circuits using ABM, the E component will be used when an output voltage is required, while the G component will be used when an output current is needed. The general form for the E extension is

E*<name> <connecting nodes>* <ABM *keyword>* <ABM *function>*

where:
   <name> is the component name assigned to the E device.
   <connecting nodes> specifies the "+ node" and the "– node", between which the component is connected.
   <ABM *keyword>* specifies the form of the transfer function to be used. There are five functions:

VALUE	Arithmetic expression
TABLE	Lookup table
FREQ	Frequency response
LAPLACE	Laplace transform
CHEBYSHEV	Chebyshev filter characteristics

   <ABM *function>* specifies the transfer function as a formula or a lookup table.

The G extension follows the same general form as the E extension. It must be noted that the PSpice current-controlled current source (or F component) and the current-controlled voltage source (or H component) do not support the analogue behavioural modelling.

Each of the ABM functions will now be described in detail.

8.3 Value

In many applications, the operation of many analogue circuits can be described by arithmetic equations. In this case, the VALUE function offers the best option of modelling. The general form of the VALUE function is

$$VALUE=\{<expression>\}$$

where <expression> can contain arithmetic functions, along with the arithmetic operators (+,-,*,/). There are a number of built-in functions available in PSpice, as shown in Table 8.1. The <expression> can also include constants, node voltages and currents, and the parameter TIME. This variable is the PSpice internal sweep variable used in the transient analysis.

To illustrate how behavioural models of analogue circuits are developed using the VALUE function, a number of examples are considered.

Example 8.3.1 Summing Amplifiers
The function of an ideal summing voltage amplifier (Figure 8.1) is simply described as

$$Esum\ 3\ 0\ VALUE=\{V1(1,0)+V2(2,0)\}$$

Table 8.1   PSpice Arithmetic Functions

Function	Meaning	Comment
ABS(x)	$\lvert x \rvert$	Absolute value of x
SQRT(x)	$x^{1/2}$	Square root of x
EXP(x)	e^x	e to the x power
LOG(x)	ln (x)	Log base e of x
LOG10(x)	log (x)	Log base 10 of x
PWR(x,y)	$\lvert x \rvert^y$	Absolute value of x to the y power
SIN(x)	sin (x)	(x in radians)
COS(x)	cos (x)	(x in radians)
TAN(x)	tan (x)	(x in radians)
ARCTAN(x)	tan^{-1} (x)	Result in radians

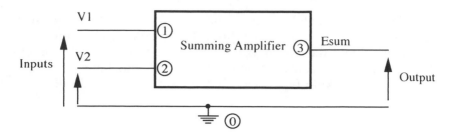

Figure 8.1   Block diagram of Example 8.3.1.

This describes a summing amplifier with an output voltage (Esum) at node 3. The value of this voltage is equal to the summation of the voltages at nodes 1 and 2. Note that nodes 1, 2, and 3 have been chosen arbitrarily to show how the "block-diagram" summing amplifier model would be connected within a system.

Example 8.3.2 Full-Wave Rectifiers
The operation of a full-wave rectifier involves inverting the negative half-cycles of the input signal, in order to create a continuous series of positive half-cycles at the output. Mathematically, this is simply taking the absolute value of the input signal. Therefore, using the VALUE function, the behavioural model of an ideal full-wave rectifier is:

$$\text{Efull 2 0 VALUE}=\{ABS(V(1)\}$$

The full-wave rectified output is at node 2, and the input to the rectifier is at node 1. Note that the output is equal to the absolute value of the input.

Example 8.3.3 Half-Wave Rectifiers
A half-wave rectifer on the other hand has only the positive half-cycles of the input signal available at the output. To develop a behavioral model of this circuit, a combination of exponential and absolute value functions are needed as shown:

$$\text{Ehalf 2 0 VALUE}=\{V(1)*(EXP(10*V(1))/EXP(ABS(10*V(1))))\}$$

Ehalf is a half-wave rectifier with the output at node 2 and the input at node 1. The number 10 in the above expression is chosen so that when the input voltage is negative, the output of the rectifier is very small or zero.

Example 8.3.4 Voltage-Controlled Oscillators (VCOs)
The output frequency of a VCO depends on the input control voltage. When a voltage, V, is applied to the input of the VCO, the output has frequency, Fo, given by:

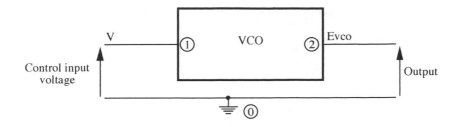

Figure 8.2   Block diagram of Example 8.3.4.

$$Fo = Fc + K * V$$

where Fc is the center frequency, K is the sensitivity of the VCO in HzV^{-1}, and V is the control input voltage. The behavioural model of an ideal VCO circuit (Figure 8.2) is:

Evco 2 0 VALUE={A*sin(twopi*TIME*(Fc+K*V(1,0)))}

Evco is a VCO with the output at node 2 and the input at node 1. The paramter A is the amplitude of the VCO, twopi is a constant, and TIME is the PSpice internal sweep variable used in the transient analysis. As an example, if A = 2V, Fc = 50 kHz, and K = 25 kHz^{-1} and the input control voltage changes, for example, from +1V to -1V, the output frequency will change from 75 kHz to 25 kHz as shown in Figure 8.3. The PSpice input file of the VCO circuit is given in Listing 8.1. The constant (twopi) in the VCO model has been defined using the .PARAM statement, and the control input signal (V) was specified using the piece-wise linear (PWL) waveform (see listing). This waveform was defined using the *StmED* program (see Appendix B for more information on this program).

8.4   Table

This function allows circuit operation to be described by a lookup table. The general form of the TABLE function is

TABLE {*expression*}=<<*input value*>, <*output value*>>

The keyword TABLE indicates that this controlled source has a tabular description. The input to the table is <*expression*>, which is evaluated, the value of which is used to look up an entry in the table. The table itself consists of pairs of values. The first value in each pair is an input, and the second is the corresponding output. Note that the table's entries must be in order from lowest to highest.

Listing 8.1   Input File of Example 8.3.4

```
VCO Block-Diagram (Figure 8.2)
*
V 1 0
+ PWL (0.000000E+00      1.000000E+00
+      0.1000000E-03    1.000000E+00
+      0.1000100E-03    0.000000E+00
+      0.2000000E-03    0.000000E+00
+      0.2001000E-03   -1.000000E+00
+      0.3000000E-03   -1.000000E+00
+      0.3001000E-03    0.000000E+00)   ; Control input signal
*
.PARAM twopi= 6.283, A = 2, Fc = 50 KHz, K = 25 KHz
*
*VCO model
*
Evco 2 0 VALUE={A*sin(twopi*TIME*(Fc + K*V(1,0)))}
*
.TRAN 0.1us 300us 0s 0.1us        ; Transient analysis range
*
.Probe V(1), V(2)                 ; Graphic output
*
.END
```

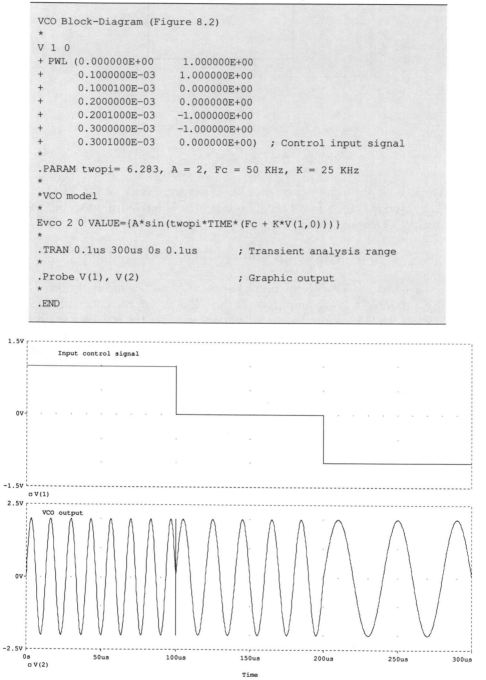

Figure 8.3   Input and output waveforms of Example 8.3.4.

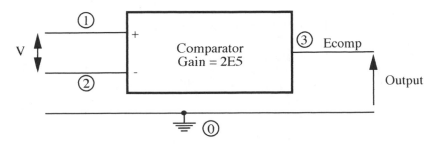

Figure 8.4 Block diagram of a comparator circuit.

Some analogue circuit operations are best described by a lookup table, an example of which is the comparator circuit.

Example 8.4.1 Comparator Circuit
Consider a comparator with the following characteristics:

$$(A>B, 1), (A<B, 0)$$

where A is the noninverting input signal, B is the inverting input signal, and 1 or 0 are the outputs of the comparator. Using the TABLE function, the behavioural model of an ideal voltage comparator (Figure 8.4) is:

Ecomp 3 0 TABLE {V(1,2)*2E5}=(0V, -15V) (0.01V, +15V)

This describes a voltage comparator with an output voltage (Ecomp) at node 3, with the noninverting and inverting inputs at nodes 1 and 2, respectively. The input to the table is the expression (V(1,2)*2E5) where V(1,2) is the comparator differential input voltage and 2E5 is the comparator gain. This expression is evaluated, the value of which is used to look up on entry in the table. In this case, the table has two pairs of values, (0V, -15V) and (0.01V, +15V), which means that if the evaluation result is ≤0, the output is -15V, while if it is ≥0.01V, the result is +15V. The number 0.01V has been chosen arbitrarily.

To illustrate the use of the comparator behavioural model, assume that a sinusoidal waveform of 10V, and 1 kHz frequency is connected to the positive input of the comparator (Figure 8.4), and a +6V DC source is connected to the comparator negative input. The PSpice net list of the comparator circuit is given in Listing 8.2. The simulated input and output waveforms of the comparator are given in Figure 8.5.

Although the TABLE function has been used in this example to describe the operation of an ideal circuit, it is well suited for modelling practical components, the performance of which can be expressed in terms of a set of measured data. For instance, the PSpice user's manual[1] gives examples of modelling the (I-V) characteristics of a tunnel diode.

Listing 8.2 Input File of Example 8.4.1

```
Comparator Block-diagram (Figure 8.4)
*
V1 1 0 sin (0V 10V 1kHz) ; sine wave input signal
V2 2 0 DC +6V              ; 6V DC signal
*
Ecomp 3 0 TABLE
+    {V(1,2)*2E5}=(0V,-15V) (0.01V, +15V); comparator model
*
.TRAN 0.01ms 2ms 0s 0.01ms    ; transient analysis range
*
.PROBE V(1), V(2), V(3)       ; graphic output
*
.END
```

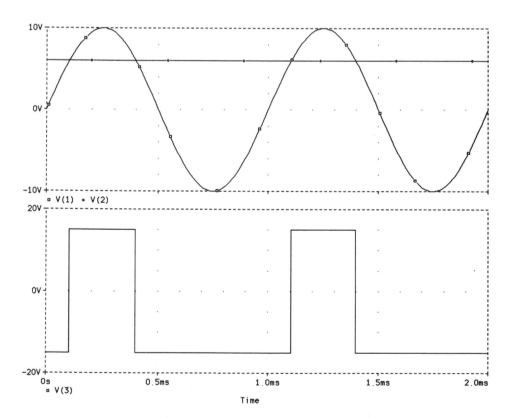

Figure 8.5 Input and output waveforms of Example 8.4.1.

8.5   Freq

This function allows the circuit operation to be described by a frequency response table. The general form of the FREQ function is

FREQ *{expression}=*
+ *<<frequency value>,<magnitude value>,<phase value>>*

The keyword FREQ indicates that this controlled source is described by a table of frequency response. The input to the table is the value of the *<expression>*. Each entry in the table consists of a frequency (in Hertz), magnitude (in decibels), and phase (in degrees). Note that the table's frequencies must be in order from lowest to highest. Interpolation is performed between entries, and for frequencies outside the table's range, zero magnitude is used.

The FREQ function differs from the TABLE function in that a component or a circuit must be described in terms of frequency response points; in the TABLE function, a component is described in terms of any (x,y) values. To demonstrate the use of the FREQ function, consider the following description of a lowpass filter.

Example 8.5.1 Filters
The normalised gain and phase response of a second-order Butterworth filter is shown in Figure 8.6. Using the FREQ function, the PSpice behavioural model of this filter is

Efilter 2 0 FREQ {V(1,0)}=(0.1,0,–10) (0.5,–0.27,–44)
+ (0.6,–0.53,–54) (0.7,–0.95,–63)
+ (0.8,–1.5,–73) (1,–3,–90)
+ (1.25,–5,–108) (2,–12,–137)
+ (3,–19,–151) (4,–24,–160)

This describes a normalised (i.e., –3dB frequency=1Hz) Butterworth lowpass filter with an output voltage at node 2 and an input voltage at node 1. To scale this filter to have a specific –3dB frequency, all the frequencies in the table must be multiplied by the required –3dB frequency. Also, if a different filter type is required, all the magnitude and phase values need to be changed according to the chosen filter type.

8.6   LAPLACE

As an alternative to the FREQ function, PSpice offers the LAPLACE function, which can be used to model the frequency response of circuits in terms of a transfer function in the Laplace transform variable s. The general form of the LAPLACE function is

LAPLACE *{<expression>}={<transform>}*

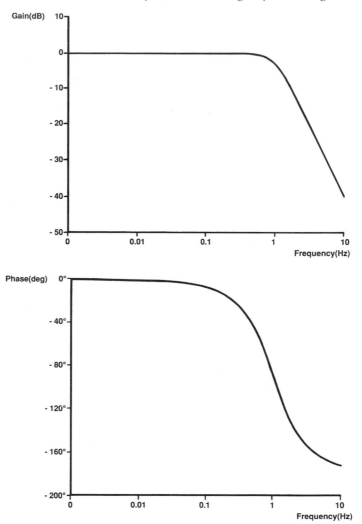

Figure 8.6 Normalised gain and phase response of a second-order Butterworth lowpass filter.

The keyword LAPLACE indicates that this controlled source has a Laplace transform description. The input to the transform is the value of *<expression>*, and *<transform>* is an expression in the Laplace variable *s*.

To illustrate the use of the LAPLACE function, consider the following description of a lowpass filter.

Example 8.6.1 Filters

The voltage transfer function of a second-order lowpass filter is

$$\frac{V_{out}}{V_{input}} = \frac{A1}{s^2 + A2s + A1} \tag{8.1}$$

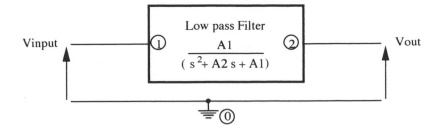

Figure 8.7 Block diagram of Example 8.6.1.

where s is the Laplace transform variable, and $A1$ and $A2$ are expressions describing the filter characteristics and are given by

$$A1 = \left(2\pi\omega_0 F_c\right)^2 \qquad A2 = \frac{2\pi\omega_0 F_c}{Q} \qquad (8.2)$$

The parameter ω_0 is the normalised frequency (in radians per second), Q is the quality factor, and F_c is the denormalised cutoff frequency of the filter. Using the LAPLACE function, the behavioural model of a second-order lowpass filter based on Equation 8.1 is

Elowpass 2 0 LAPLACE {V(1,0)}={A1/(s*s+A2*s+A1)}

where the input to the filter (Fig. 8.7) is a voltage at node 1, and the output is a voltage at node 2.

The simple nature of the transfer function allows changes in the model to be made easily. For example, changing the filter type simply involves entering the appropriate values of the parameters ω_0 and Q selected from Table 8.2 into the above model.

Figure 8.8 shows the frequency response (gain and phase) of a 5kHz, 0.5dB ripple Chebyshev second-order lowpass filter. The input file of this filter is given in Listing 8.3. Although the listing shows that the AC (frequency) analysis is performed on a linear scale, the graphs of Figure 8.8 have been plotted on a log scale. This is achieved by selecting the "X-axis" option from the *Probe* menu, and then "Log" from

Table 8.2 ω_0, Q of the Various Filter Responses

	ω_0	Q
Butterworth	1	0.707
Chebyshev (0.5dB)	1.231	0.864
Chebyshev (1dB)	1.05	0.957
Chebyshev (2dB)	0.907	1.129
Chebyshev (3dB)	0.841	1.307
Bessel	1.73	0.577

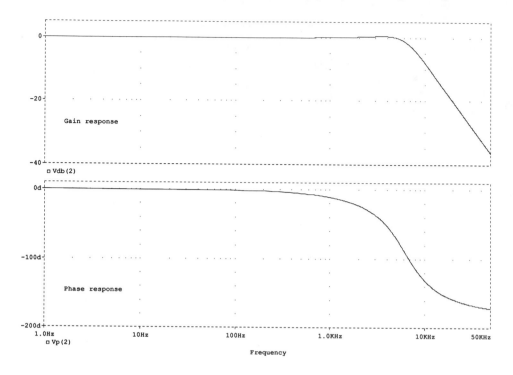

Figure 8.8   Frequency response (gain and phase) of Example 8.6.1.

Listing 8.3   Input File of Example 8.6.1

```
0.5dB ripple, 5kHz Chebyshev 2nd-order lowpass filter
*
Vinput 1 0 AC 1
*
.AC LIN 500 1 50K                   ; linear frequency range
*
.PARAM omega={1.231}, Q={0.864}   ; see Table 8.2
*
.PARAM Fc={5kHz}                    ; -3dB frequency point
*
.PARAM pi=3.14, pi_square={pi*pi}         ; constants
*
.PARAM A1={4*pi_square*omega*omega*Fc*Fc}  ; see Eq.(8.2)
.PARAM A2={(2*pi*omega*Fc)/Q}              ; see Eq.(8.2)
*
* Filter transfer function (Eq.8.1)
*
Elowpass 2 0 LAPLACE {V(1,0)}={(A1)/(s*s+s*A2+A1}
*
.PROBE V(2)                            ; filter output
*
.END
```

the submenu. Note that it is possible to obtain sharper roll-off filters by cascading a number of these second-order filter models.

So the functions LAPLACE and FREQ can both be used to model frequency response, but clearly, as can be seen from the filter examples (8.5.1, and 8.6.1), the LAPLACE function offers more flexibility with respect to circuit modifications but at the expense of developing and using a transfer function. Models of other filter functions such as highpass and bandpass are easily developed using the LAPLACE function[3].

8.7   Chebyshev

This function allows the user to simulate Chebyshev filters directly. In other words no transfer functions or frequency response tables are required. The general form of the CHEBYSHEV function is

CHEBYSHEV {<*expression*>} = + <*type*>, <*cutoff frequencies*>,
<*attenuations*>

The keyword CHEBYSHEV indicates that this controlled source is described in terms of specification for a Chebyshev filter. The input to the filter is the value of <*expression*>. There are four filter types available: lowpass (LP), highpass (HP), bandpass (BP), and band reject (BR). Figure 8.9 shows the frequency characteristics of the various filter types.

Note that there are two cutoff frequencies (Fp, and Fs) in the case of LP and HP filters, and four cutoff frequencies (Fs1, Fp1, Fp2, and Fs2) in the case of BP and BR filters. The cutoff frequencies define the boundaries of the filter passband(s) and stopband(s). Two attenuation values follow the cutoff frequencies. These define the maximum allowable attenuation (A_{max}) in the passband and the minimum required attenuation (A_{min}) in the stopband (both values are in decibels). Over the years, many good textbooks have been written on analogue filter design.[4-6]

To illustrate the use of the CHEBYSHEV function, consider the following examples.

Example 8.7.1 Lowpass Filters

Elowpass 2 0 CHEBYSHEV {V(1,0)}=LP 20kHz 22kHz 0.1dB 60dB

This describes a Chebyshev lowpass filter (Figure 8.10) with the following characteristics: passband edge is 20kHz, passband attenuation is 0.1dB, stopband edge is 22kHz, and stopband attenuation is 60dB. The input to the filter is a voltage at node 1, and the output is a voltage at node 2.

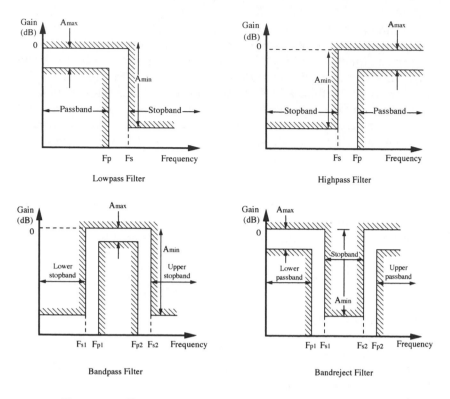

Figure 8.9   Frequency characteristics of various filter types.

Example 8.7.2 Highpass Filters

Ehighpass 2 0 CHEBYSHEV {V(1,0)}=HP 22kHz 20kHz 0.1dB 60dB

This expression models the same filter as in Example 8.7.1 but with highpass characteristics.

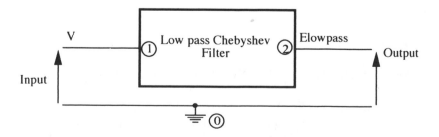

Figure 8.10   Block diagram of Example 8.7.1.

Example 8.7.3 Bandpass Filters

Ebandpass 2 0 CHEBYSHEV {V(1,0)}= +BP 1kHz 3kHz 30kHz
60kHz 0.1dB 60dB

This is a bandpass filter with the passband between 3 and 30kHz and the stopband ≤1 and ≥60kHz. The passband ripple is 0.1dB, and stopband attenuation is 60dB.

Example 8.7.4 Band Reject Filters

Enotch 2 0 CHEBYSHEV {V(1,0)}= +BR 1kHz 3kHz 30kHz 60kHz
0.1dB 60dB

This is a band reject or notch filter with passband of ≤1 and ≥60kHz and stopband between 3 and 30kHz.

The usefulness of the Chebyshev function is limited when compared with the FREQ and LAPLACE functions because

1. It is restricted to describing filters with Chebyshev characteristics.
2. It does not indicate to the user what the filter order (circuit complexity) will be. This information is important, since when the designer is satisfied with the block-diagram system simulation, the circuit-level design is assumed to follow. To complete the circuit-level design, the order of the filter must be known.

8.8   Application example

In this section, some of the behavioural circuit models developed earlier will be used to simulate an AM modulator and demodulator. These circuits have been selected because PSpice does not provide an AM modulation directly. They also strike a reasonable balance of complexity to illustrate the usefulness of ABM while still allowing a comparison with the primitive level. Figure 8.11 shows the block diagram of the AM system. It consists of a summing amplifier and an analogue multiplier, which describe the modulator, and a rectifier and lowpass filter, which represent the demodulator.

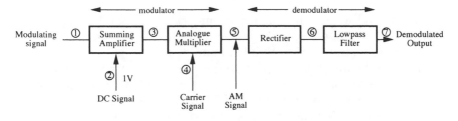

Figure 8.11   Block diagram of an AM system.

Listing 8.4   Input File of an AM System

```
AM system (Figure 8.11)
*
Vmod 1 0 sin (0 0.5V 1kHz 0 0 90) ; Modulating signal
*
Vdc 2 0 DC 1V                      ; DC input voltage
*
Vcar 4 0 sin (0 1V 100kHz)         ; Carrier signal
*
Esum 3 0 VALUE={V(1)+V(2)}         ; Summing amplifier model
*
Emult 5 0 VALUE={V(3)*V(4)}        ; AM output signal
*
*Half-Wave Rectifier Model
Erect 6 0 VALUE={V(5)*(EXP(10*V(5))/EXP(ABS(10*V(5))))}
*
*2nd-Order Butterworth Lowpass Filter Model
*
.PARAM Q={0.707}, omega={1}        ; see Table 8.2
*
.PARAM Fc={1kHz}                   ; -3dB frequency
*
.PARAM pi={3.14}, pi_square={PI*PI} ; Constants
*
.PARAM A1={4*pi_square*omega*omega*Fc*Fc}   ; see Eq.(8.2)
*
.PARAM A2={(2*pi*omega*Fc)/Q}               ; see Eq.(8.2)
*
Elow 7 0 LAPLACE {V(6)}={(A1)/(s*s+s*A2+A1)}; Filter model
*
.TRAN 0.1ms 2ms 0ms 0.1ms          ; Transient analysis range
*
.Probe V(1),V(5),V(7)   ; Modulating, AM & demodulated signal
*
.END
```

An AM modulated signal is described by

$$V_{am} = A \sin \omega_c t (1 + m \cos \omega_m t) \qquad (8.3)$$

where ω_m and ω_c are the angular frequency of the modulating and carrier signals, respectively, A is the amplitude of the carrier, and m is the modulation index. The modulation index is defined as the ratio of the peak of the modulating signal to the peak of the carrier signal.

Examination of the above equation suggests that an AM signal can be modelled in two stages using ABM. In the first stage, the modulating signal ($m \cos \omega_m t$) is added to a fixed DC voltage of 1V to obtain ($1+ m \cos \omega_m t$). In the second stage, the result of this addition will be multi-

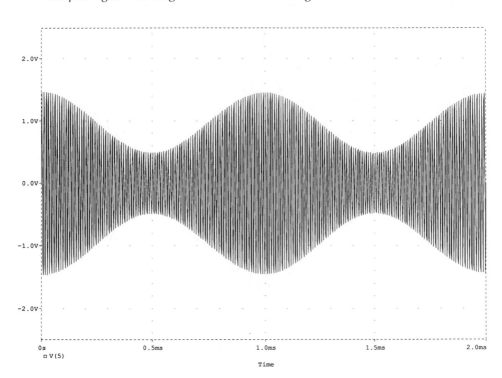

Figure 8.12 PSpice AM signal (modulating signal=0.5V, 1kHz; carrier signal=1V, 100kHz).

plied by the carrier (A sin $\omega_c t$). Both of these operations are easily performed using the VALUE function of the ABM.

Simulating the demodulator involves modelling the operations of the rectifier and the lowpass filter. Both of these circuits have already been described in the previous examples. In the case of the lowpass filter, the Laplace transfer (LAPLACE) function model (Example 8.6.1) will be chosen in preference to the frequency response table (FREQ) model (Example 8.5.1) because it offers greater flexibility. The second-order filter has been assumed to be of the Butterworth type with –3dB frequency at 1kHz.

The PSpice input file of the AM system is given in Listing 8.4. The modulating signal (1kHz), a DC voltage (1V), and a carrier signal (100kHz) have been defined using various independent voltage sources, as shown at the start of the PSpice input file. Here, the modulation index has been assumed to be 0.5, or 50%. Note that the modulating signal (cos $\omega_m t$) has been described using a sine wave source with a 90° phase shift, since cos ωt=sin(ωt+90). Figures 8.12 and 8.13 show PSpice simulation of the AM signal and its frequency spectrum.

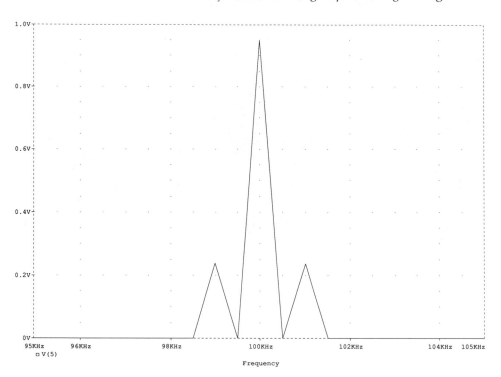

Figure 8.13   Frequency spectrum of an AM signal (modulating signal=1kHz; carrier signal=100kHz).

PSpice correctly predicts the carrier at 100kHz and the two sidebands at 99 and 101kHz, respectively. Note that Figure 8.12 was obtained by typing "V(5)", having selected "Add_trace" from the *Probe* menu. Having done this, selecting "Fourier" from the "X-axis" menu generates Figure 8.13.

Figure 8.14 shows the demodulated signal at the filter output, which corresponds to the input signal but with a reduced amplitude due to the filtering function at the demodulator.

8.8.1   Primitive-level approach

As a comparison, the same AM system was simulated using primitive level components, as shown in Figure 8.15. The summing amplifier was simulated using an ideal VCVS, while the multiplier circuit was simulated using a four quadrant Gilbert Cell[7] based on ideal transistors and current sources. For simplicity, the multiplier circuit is shown as a block diagram in Figure 8.15. The rectifier was simulated using a combination of a diode, a resistor, and a capacitor. The filter was modelled

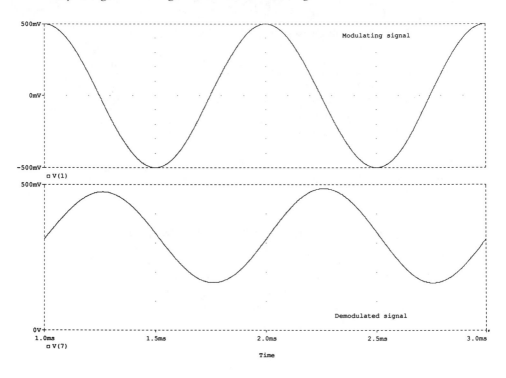

Figure 8.14 Modulating and demodulated signals of Example 8.8.

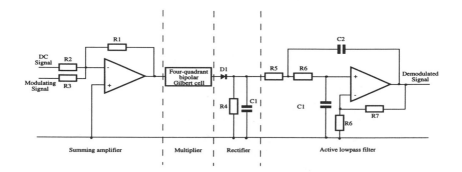

Figure 8.15 Primitive-level circuit of an AM system.

using a VCVS, four resistors, and two capacitors and was implemented using an active Sallen-Key circuit[4]. Both behavioural and primitive approaches have been shown to yield similar and satisfactory results. However, examination of Table 8.3 indicates a saving in development

Table 8.3 Comparison of the Two Simulation Approaches

	Behavioural	Primitive
Development time (h)	4	7
Simulation time (s)	175	220
Number of system nodes	7	26

Note: Simulation time was measured on a PC 386SX.

and simulation time associated with the behavioural approach. Clearly optimisation of system performance is most effectively achieved by the use of ABM.

8.9   Simulation of integration and differentiation

Although the built-in PSpice arithmetic functions (Table 8.1) are comprehensive, it does not contain all the functions that may be required. Two examples are differentiation and integration, both of which are extensively used in electronic circuits. To illustrate the integration process, consider the following equation:

$$vo = \frac{-1}{R1C1} \int_0^t vi(t)dt \qquad (8.4)$$

This shows the output voltage (*vo*) is the (inverted) integral of the input voltage (*vi*), multiplied by the constant (1/*R1C1*). The voltage (*v*) across a capacitor (*C*) is given by:

$$v = \frac{1}{C} \int_0^t i(t)\,dt \qquad (8.5)$$

Comparing the two equations shows that the term *(−vi/R1)* can be described using a voltage-controlled current source (i.e., the G component). The output current of this source flows into the capacitor (*C1*), and the voltage across this capacitor represents the output voltage (*vo*). This description is shown in Figure 8.16.

As an example, assume that the input signal is a voltage pulse of 1V amplitude, rise time and fall time of 10ns, and a period of 1ms. Also, assume that the values of *R1* and *C1* are 1kΩ and 0.1µF, respectively. The input file of Equation (8.4) is given in Listing 8.5. An ideal VCCS has an infinite input impedance and an infinite output impedance; therefore, resistors (RA and RB) must be included in the listing to satisfy PSpice requirements of floating nodes. The values of these resistors are chosen to be large enough to have no effect on the circuit operation. The simulated input and output waveforms of the integration model (Equation [8.4], and Figure 8.16) are given in Figure 8.17.

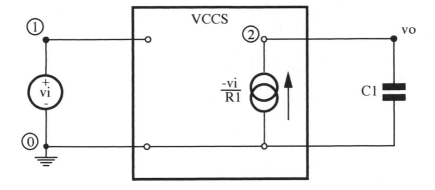

Figure 8.16 Integration model.

Listing 8.5 Input File of Integration Model (Figure 8.16)

```
Integration Model  (Figure  8.16)
*
Vi  1  0  pulse  (0V  1V  0s  10ns  10ns  0.5ms  1ms) ;input  signal
*
.TRAN  0.01ms  3ms  0ms  0.01ms ;  transient  analysis  range
*
.PARAM  C1=0.1uF,  R1=1k          ;  component  values
*
G  0  2  VALUE={-1*V(1,0)/R1}  ;  describes  a  VCCS  value  of  -vi/R1
*
C1  2  0  {C1}      ;  voltage  across  C1  gives  vo
*
RA  1  0  1G        ;  resistor  required  to  satisfy  PSpice  rules
*
RB  2  0  1G        ;  resistor  required  to  satisfy  PSpice  rules
*
.PROBE
*
.END
```

These waveforms are as expected from an ideal active integrator circuit (Figure 8.18), with the input and output voltage relationship described in Equation (8.4).

To illustrate how differentiation is simulated in PSpice, consider the following equation:

$$vo = R1C1\frac{dvi}{dt} - R1C1\frac{dvo}{dt} \tag{8.6}$$

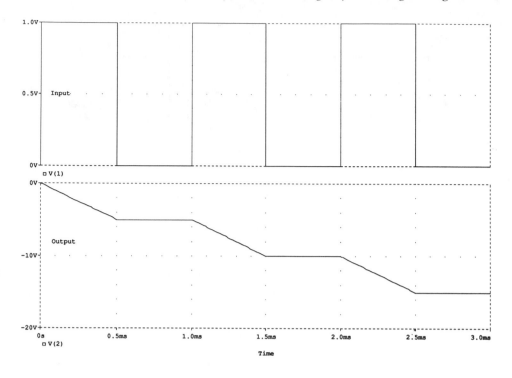

Figure 8.17   Input and output waveforms of Equation (8.4) and Figure 8.16.

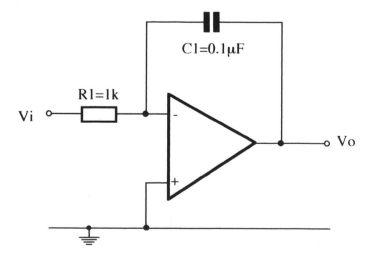

Figure 8.18   Ideal active integrator.

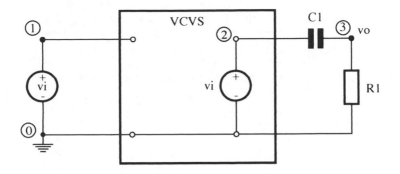

Figure 8.19 Differentiation model.

The current (i) through a capacitor (C) is given by

$$i = C\frac{dv}{dt} \qquad (8.7)$$

Rewriting Equation (8.6) as

$$\frac{vo}{R1} = C1\frac{d}{dt}(vi - vo)$$

and comparing with Equation (8.7) shows that

$$i = \frac{vo}{R1} \qquad (8.8)$$

$$v = (vi - vo) \qquad (8.9)$$

To produce Equation (8.8), a resistor is required across the output. To produce the voltage across the capacitor ($C1$), a voltage-controlled voltage source (i.e., the E component) must be connected to one end of the capacitor with a value equal to vi, and the other end of the capacitor should be connected to the output (vo) as shown in Figure 8.19.

As an example, assume that the input signal is a voltage pulse of 1V amplitude, rise time and fall time of 10ns, and a period of 1ms. Also, assume that the values of $R1$ and $C1$ are 1kΩ and 0.1μF, respectively. The input file of Equation (8.6) is given in Listing 8.6. An ideal VCVS has an infinite input impedance and zero output impedance; therefore, a resistor (RA) across the VCVS input must be included in the listing to satisfy PSpice requirements of floating nodes. The value of this resistor is chosen to be large enough to have no effect on the circuit operation. The simulated input and output waveforms of the differentiation model (Equation [8.6] and Figure 8.19) are given in Figure 8.20.

Listing 8.6   Input File of Differentiator Model (Figure 8.19)

```
Differentiation Model (Figure 8.19)
*
Vi 1 0 pulse (0V 1V 0s 10ns 10ns 0.5ms 1ms) ; Input pulse
*
.TRAN 0.01ms 3ms 0ms.01ms        ; Transient analysis range
*
.PARAM R1=1K, C1=0.1uF           ; Component values
*
E  2  0 VALUE={V(1,0)}           ; VCVS with value of Vi
*
C1 2 3 0.1uF
*
R1 3 0 1K
*
RA 1 0 1E9      ; Resistor required to satisfy PSpice rules
*
.PROBE
*
.END
```

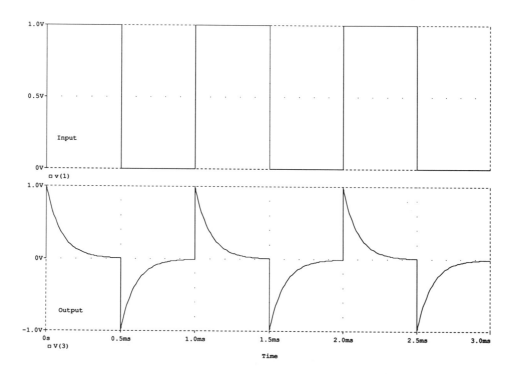

Figure 8.20   Input and output waveforms of Equation (8.6) and Figure 8.19.

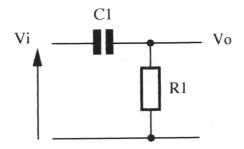

Figure 8.21   Simple differentiator circuit.

These waveforms are as expected from a simple differentiator circuit (Figure 8.21), with the input and output voltage relationship described in Equation (8.6).

The ability of PSpice to simulate differentiation and integration using ABM allows the designer to model electromechanical systems such as relays and DC motors. This is because the operations of most mechanical systems are often described by either integration, differentiation, or both. Behavioral models of relays and DC motors are part of the MISC.LIB of the PSpice library. For a detailed discussion on the modelling of relays and DC motors, see Reference 8.

8.10 Modelling practical components

Most of the examples given up to this point have dealt with modelling ideal components and circuits. ABM can be used to model practical components. As an example, consider the modelling of the gain-frequency characteristics of an op-amp.

Example 8.10.1 Op-amp Modelling

The ideal model of an op-amp (i.e., a VCVS) makes no allowances for the gain-bandwidth product parameter in practical op-amps. Therefore, an amplifier circuit will have the same gain at DC and at infinite frequency. Clearly, it is undesirable to use this model in many applications, and a more representative model is needed. A comprehensive model is given in the PSpice library based on the standard Boyle op-amp model[2]. This model is discussed in Chapter 6, Section 6.5. Sometimes the full Boyle op-amp model is not required, and ABM provides an elegant way of modelling the gain-bandwidth product and phase characteristics of op-amps. For a general-purpose op-amp, such as the 741, the open loop gain transfer function is given by

$$Av(s)=A_0/(1+s/2\pi F_p)$$

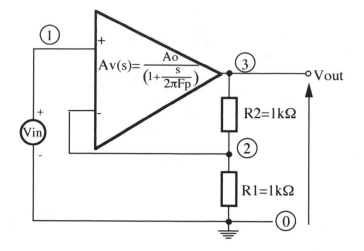

Figure 8.22   Amplifier circuit based on op-amp analogue behavioural model.

where A_0 is the low frequency value of the open loop gain, F_p is the –3dB frequency of the open loop gain of the op-amp, and s is the Laplace variable. This transfer function assumes a one-pole roll-off op-amp model[4]. This means that the gain characteristic of the op-amp has a slope of –20dB per decade and a phase shift of –45° at the –3dB fre-

Listing 8.7   Input File of Example 8.10.1

```
Amplifier Circuit (Figure 8.22)
*
Vin 1 0 AC 1
*
.AC LIN 500 1 1E6
*
.PARAM Ao={200K}      ; 741 op-amp low frequency open loop gain
*
.PARAM Fp={5Hz}       ; 741 op-amp –3dB frequency point
*
.PARAM PI=3.1416      ; Constant
*
R1  2  0  1K
*
R2  2  3  1K          ; Amplifier gain resistor
*
Eop-amp 3 0 LAPLACE
+            {V(1,2)}={Ao/(1+s/(2*PI*Fp))} ; op-amp model
*
.PROBE V(3)                               ; Amplifier output
*
.END
```

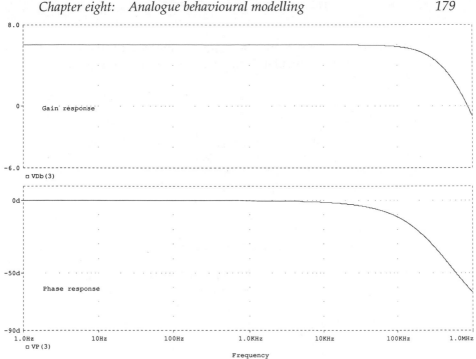

Figure 8.23 Frequency response of the circuit in Figure 8.22.

quency. It is possible to produce transfer functions for multiple poles and zeros of op-amps[4,7].

To illustrate the use of this model, consider the simulation of the amplifier circuit shown in Figure 8.22. This amplifier has a gain of 2. The op-amp open loop gain has been written in terms of s, allowing the LAPLACE function to be used. Also, since the output of the op-amp is a voltage, the E component will be used. The input file of the amplifier circuit is given in Listing 8.7. It has been assumed that a 741 op-amp with a gain-bandwidth product of 1MHz and A_0 of 200E3 (yielding F_p=5Hz) will be used. Figure 8.23 shows the simulated frequency response (gain and phase) of the amplifier circuit. The –3dB point is at 500kHz, as expected for a 741 with a gain of 2. The amplifier circuit has a phase shift of –45° at the –3dB frequency point.

8.11 Chapter summary

Analogue behavioural modelling (ABM) allows the designer to use a hierarchy simulation approach. At the system level, it enables the designer to simulate systems as a combination of block diagrams, each of which is modelled using a mathematical equation or data table. At the circuit level, it allows models of an appropriate complexity to be developed.

Its strength is in allowing the designer to model with the level of accuracy needed at any particular design stage with a minimum amount of simulation.

References

1. MicroSim Corporation, *Circuit Analysis — User's Guide Manual* (The Design Center), Version 5.3, Irvine, CA, January 1993.
2. Boyle, G.R., Cohn,B.M., Pederson, D., and Solomon, J.E., "Macromodelling of Integrated Circuit Operational Amplifiers", *IEEE Journal of Solid-State Circuits*, Vol. SC-9, 1974, pp. 353-364.
3. Al-Hashimi, B., "Behavioral Model Emulates Universal Filter", *EDN Magazine*, February 4, 1992, pp. 124.
4. Van Valkenburg, M.E., *Analogue Filter Design*, Holt, Rinehart & Winston, New York, 1982.
5. Williams, A.B., *Electronic Filter Design Handbook*, McGraw-Hill, 1981.
6. Stephenson, F.W., *RC Active Filter Design Handbook*, John Wiley & Sons, New York, 1985.
7. Gray, P.R. & Meyer, R.G., *Analysis and Design of Analogue Integrated Circuits*, John Wiley & Sons, New York, 1984, chapter 10 (pp. 590).
8. MicroSim Corporation, *Application Notes Manual* (The Design Center), Version 5.3, Irvine, CA, January 1993.

chapter nine

Digital and mixed analogue/digital simulations

This chapter introduces the basics of digital and mixed analogue/digital circuit simulation. It includes a discussion on PSpice digital primitive and stimulus devices. The chapter also reviews the PSpice digital components library. Examples of digital and mixed analogue/digital circuit simulations are given to illustrate the various steps, which include component entry, digital input, waveform generation, and analysis.

9.1 Digital components description

PSpice recognises the digital components shown in Table 9.1. To describe a digital component, three statements are required. The first, a digital primitive description statement, defines the component type with input and output nodes. The second, a timing model statement, describes the component timing characteristics such as propagation delays, setup, and hold times. The third, an input/output model statement, describes the component loading and driving characteristics, as well as interfacing nodes, which are discussed in Section 9.6.

9.1.1 Description of Logic Gates

PSpice supports two types of combinational gates: standard and tri-state, as shown in Table 9.1. The basic form of the standard gate digital primitive description statement is

> U<name> <type> [(number of inputs)]
> + <$D_DPWR> <$D_DGND> <input nodes> <output nodes>
> + <timing model name> <I/O model name>

where U is the PSpice symbol for a digital primitive device, and <name> is the gate name, which can be up to eight characters long. The name

Table 9.1 PSpice Digital Components

Component	Type	Meaning
Standard gates	BUF	Buffer
	INV	Inverter
	AND	AND gate
	NAND	NAND gate
	OR	OR gate
	NOR	NOR gate
	XOR	Exclusive OR gate
	NXOR	Exclusive NOR gate
	BUKFA	Buffer array
	INVA	Inverter array
	ANDA	AND gate array
	NANDA	NAND gate array
	ORA	OR gate array
	NORA	NOR gate array
	NXORA	Exclusive OR gate array
	AO	AND-OR compound gate
	AOI	AND-NOR compound gate
	OAI	OR-NAND compound gate
Tri-state gates	BUF3	Buffer
	INV3	Inverter
	AND3	AND gate
	NAND3	NAND gate
	OR3	OR gate
	NOR3	NOR gate
	XOR3	Exclusive OR gate
	NXOR3	Exclusive NOR gate
	BUF3A	Buffer array
	INV3A	Inverter array
	AND3A	AND gate array
	NAND3A	NAND gate array
	OR3A	OR gate array
	NOR3A	NOR gate array
	XOR3A	Exclusive OR gate array
	NXOR3A	Exclusive NOR gate array
Transfer gate	NBTG	N-channel transfer gate
	PBTG	P-channel transfer gate
Flip-flops and latches	JKFF	J-K, negative-edge triggered
	DFF	D-type, positive-edge triggered
	SRFF	S-R latch
	DLTCH	D-type latch
Pullup and pulldown resistors	PULLUP	Pullup resistor array
	PULLDN	Pulldown resistor array
Delay lines	DLYLINE	Delay line

Table 9.1 (continued)   PSpice Digital Components

Component	Type	Meaning
Programmable logic	PLAND	AND array
arrays	PLOR	OR array
	PLXOR	Exclusive OR array
	PLNAND	NAND array
	PLNOR	NOR array
	PLNXOR	Exclusive NOR array
	PLANDC	AND array, true and complement
	PLORC	OR array, true and complement
	PLXORC	Exclusive OR array, true and complement
	PLNANDC	NAND array, true and complement
	PLNORC	NOR array, true and complement
	PLNXORC	Exclusive NOR array, true and complement
Memory	ROM	Read-only memory
	RAM	Random-access memory
A/D and D/A	ADC	Multibit A/D converter
converters	DAC	Multibit D/A converter

can contain either letters, numbers, or both. Examples of digital primitive names are U1 and U5a.

The parameter *<type>* defines the type of standard gate given in Table 9.1. For example, an AND gate is identified by the type AND, while a NOR gate is identified by the type NOR.

The parameter [*(number of inputs)*] specifies the number of inputs to the gate, and *<$D_DPWR>* and *<$D_DGND>* define the digital power and ground nodes of the digital component; in PSpice they have the default values of 5 and 0V, respectively.

The parameters *<input nodes>* and *<output nodes>* specify the input and output nodes of the gate, respectively.

The parameter *<timing model name>* is the name of the model that describes the gate timing characteristic. This is similar to the model name used in the semiconductor components discussed in Chapter 2, Section 2.3. The model name can begin with any character and can be up to eight characters long. Examples are T1 and T_NOR. The basic form of the .MODEL timing model statement of the standard gate is

.MODEL <timing model name> UGATE [model parameters]

where the parameter *<timing model name>* is the name given to the gate in the digital primitive device description statement. The parameter

Table 9.2   Standard Gate Timing Model Parameters

Model parameter	Description	Default	Unit
TPLHMIN	Delay: low to high, min	0	s
TPLHTY	Delay: low to high, nom	0	s
TPLHMX	Delay: low to high, max	0	s
TPHLMN	Delay: high to low, min	0	s
TPHLTY	Delay: high to low, nom	0	s
TPHLMX	Delay: high to low, max	0	s

UGATE is the PSpice timing model symbol for a standard gate. Table 9.2 describes available timing model parameters and their default values.

These parameters are specified by the user to set the gate propagation delay, which is expressed in terms of minimum, nominal, and maximum values. For example, the nominal low-to-high propagation delay of a gate is specified as TPLHTY (see Table 9.2). If timing model parameters of a gate are not specified by the user, PSpice will set the parameters to their default values to give ideal component performance. Model parameter values are usually obtained from the manufacturer's component data sheets.

The parameter<*I/O model name*> is the name of the input/output model, which describes the gate driving characteristics and loading. The model name can begin with any character and can be up to eight characters long. Examples are IO1 and IO_AND. The basic form of the .MODEL I/O statement of a digital component is

.MODEL <I/O model name> UIO [model parameters]

where the parameter <*I/O model name*> is the name given to a component in the digital primitive description statement. The parameter *UIO* is the PSpice I/O model type for a digital component. PSpice has 20 I/O model parameters, Table 9.3 shows some of these parameters; for a complete list see Reference 1. If I/O model parameters of a gate are not specified by the user, PSpice will set the parameters to their default values to give ideal component performance.

For an illustration of the description of gates, consider the circuit in Figure 9.1. The PSpice description of the circuit is

```
U1  NAND(2)  $D_DPWR  $G_DGND  A  B  X  T1  IO1
U2  NOR(2)   $D_DPWR  $G_DGND  C  X  Y  T1  IO1
.MODEL  T1  UGATE
.MODEL  IO1  UIO
```

Table 9.3 Some Input/Output Model Parameters

Model parameter	Description	Default	Unit
INLD	Input load capacitance	0	F
OUTLD	Output load capacitance	0	F
DRVH	Output high-level resistance	50	O
DRVL	Output low-level resistance	50	O

The AND gate, U1, has its two inputs at nodes A and B, and its output at node X. The NOR gate, U2, has its two inputs at nodes C and X, and its output at node Y. The G_GPWR and G_GDND are digital power and ground nodes of the respective gates. The default values of these nodes are 5 and 0V, respectively. It has been assumed that both gates have the same timing and I/O models. The names of these models, which have been chosen arbitrarily, are T1 and IO1, respectively. Note that the timing and I/O model parameters have been set to their default values, since no model parameters were specified in the .MODEL statements.

9.1.2 Flip-flops and latches

PSpice supports edge-triggered flip-flops and latches. There are two types of edge-triggered flip-flops: the J-K, which is negative edge triggered, and the D-type, which is positive edge triggered (see Table 9.1). There are two types of latches: the S-R and the D-type. Only the edge-triggered D-type flip-flop will be considered here. The basic form of the D-type flip-flop digital primitive device description statement is

> *U<name> DFF <no. of flip-flops> <$D_DPWR> <$D_DGND>*
> *+ <presetbar node> <clearbar node> <clock node>*
> *+ <D node 1 >...<D node n>*
> *+ <Q output 1>..<Q output n>*
> *+ <Qbar output 1>..<Qbar output n>*
> *+ <timing model name> <I/O model name>*

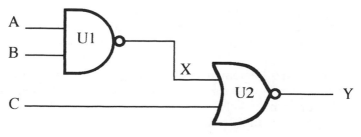

Figure 9.1 Simple logic gates circuit.

Table 9.4 Some Edge-Triggered Flip-Flop Timing Model Parameters

Model parameter	Description	Default	Unit
TPPCQLHMN	Delay:preb/clrb to q/qb low to high, min	0	s
TWPCLTY	Min preb/clrb width low, min	0	s
TSUDCLKMN	Setup: j/k/d to clk/clkb edge, min	0	s
THDCLKMN	Hold: j/k/d after clk/clkb edge, min	0	s

where *U* is the PSpice symbol for the digital primitive device, and *<name>* is the flip-flop name; example names are U2 and U3b. The parameter *DFF* is the PSpice symbol for a positive edge-triggered D-type flip-flop.

The parameter *<no. of flip-flops>* specifies the number of flip-flops, and *<$D_DPWR>* and *<$D_DGND>* define the digital power and ground nodes of the component. The specifications of the input, output, clock, and other control nodes of the flip-flop are self-explanatory. Note that the parameters *<presetbar node>*, *<clearbar node>*, and *<clock node>* are common to all the flip-flops specified.

The basic form of the .MODEL timing model statement of edge-triggered flip-flop is

.MODEL <timing model name> UEFF [model parameters]

which is of a format similar to that of the standard gate, apart from the symbol *UGATE*, which has been replaced by *UEFF* (the PSpice timing model symbol for edge-triggered flip-flops).

Some of the [*model parameters*] of the timing model of edge-triggered flip-flops are given in Table 9.4. PSpice has 30 model parameters, some of which are shown in Table 9.4. For a complete list of the model parameters, see Reference 1.

The timing model parameters are specified by the user to set the propagation delay, setup time, and hold time of the flip-flop. All these parameters are expressed in terms of minimum, nominal, and maximum values. For example, the minimum low-to-high propagation delay of a flip-flop gate is specified as TPPCQLHMN (see Table 9.4). If model parameters of the flip-flop are not specified by the user, PSpice will set the parameters to their default values to give ideal component performance.

The .MODEL I/O statement of edge-triggered flip-flops is the same as the standard gate given earlier but obviously with different values for the model parameters. Example 9.3.2 illustrates the description of positive-edge D-type flip-flops.

Table 9.5 PSpice Digital States

State	Meaning
0	Low, false, no, off
1	High, true, yes, on
R	Rising (change from 0 to 1 during rising edge) ↑
F	Falling (change from 1 to 0 during falling edge) ↓
X	Do not care, not defined
Z	High impedance

9.2 Digital waveform generation

So far, gates and edge-triggered flip-flops have been described. As discussed in Section 2.5 of Chapter 2, for analogue simulation, PSpice provides voltage- and current-independent sources, which allow the user to generate a variety of analogue input signals. There are two methods of generating digital waveforms in PSpice. The first method is the stimulus generator, or (STIM), which uses simple description statements to define the digital waveform. The second method is the file stimulus (FSTIM), which obtains the digital waveforms from an external file and can be used to provide input waveforms to a number of simulations. Here, only the stimulus generator (STIM) will be discussed since it is the method most commonly used to generate digital waveforms. Before the stimulus generator is discussed, it is necessary to explain the digital logic levels (or states) available in PSpice and their notation in *Probe*. Table 9.5 gives the PSpice digital states, and Figure 9.2 shows their notation in *Probe*.

Note that PSpice refers to logic levels and not actual voltages. For example, logic level "1" means only that the voltage is somewhere within the "high" range of a particular digital component.

9.2.1 Stimulus generator (STIM)

To generate a digital waveform using the STIM device, a digital stimulus description statement is required, and its basic form is

U<name> STIM (no. of signals, format) <$D_DPWR> <$D_DGND>
+ <node(s)> IO_STM [TIMESTEP=<stepsize>] <command>

where *U* is the PSpice symbol for the digital stimulus device, and *<name>* is the stimulus device name; example names are Uclk and Uset. The parameter *STIM* is the PSpice symbol for the stimulus generator.

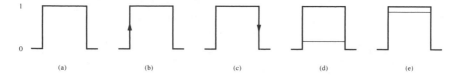

Figure 9.2 Graphical notation of various digital states in *Probe*: (a) logic levels (0,1); (b) rising edge (R); (c) falling edge (F); (d) do not care (X); (e) high impedance (Z).

The parameter (*no. of signals, format*) specifies the number of signals created by the stimulus generator and their format. There are three formats available, binary, octal, and hexadecimal, which are selected by setting <format> to the digit 1, 3, or 4, respectively. <$D_DPWR> and <$D_DGND> define the digital power and ground nodes.

The parameter <*node(s)*> defines the node names of the signals created by the stimulus generator. The number of nodes must be equal to the number of signals specified in the <*no. of signals*> parameter. The parameter *IO_STM* is the name of the input/output model, which describes the driving characteristics of the stimulus generator.

This model is part of the PSpice digital component library (see Section 9.4). The user must reference the *IO_STM* model in the circuit-input file using a .LIB command. This is achieved as follows:

.LIB C:\digital.lib

This statement assumes that the PSpice digital library is on drive C of the PC. The parameter [*TIMESTEP*] is an optional parameter that defines the pulse width of the digital waveform. This is very useful when generating repetitive waveforms such as clock signals. If *TIMESTEP* is not specified, the default is 0s.

The parameter <*Command*> defines the shape of the waveform to be generated. The basic form of the <*command*> parameter is <<*time*> <*value*>>, which specifies the digital waveform in terms of two coordinates (time and value). The waveform values could be any of the logic states shown in Table 9.5; these are 0, 1, R, F, X, or Z. The time can be specified in seconds (with the "S" suffix) or in clock steps (with the "C" suffix). Time values may be absolute, for example, 15ns or 10C, or relative to the previous time, in which case the time value must be prefixed with a "+", for example, +15ns or +10C.

PSpice offers a number of variants to the basic command (time and value) to achieve different waveforms. One example shown here allows the generation of periodic waveforms by the use of a loop command. This command has the form

<<*time*> <*value*>>
<*LABEL*>=<*label name*>
<<*time*> GOTO <*label name*> <*n*> TIMES>

This type of command, when used with the [*TIMESTEP*] parameter, can provide a very efficient method of generating repetitive waveforms such as clock signals. This is achieved by specifying the first step of the waveform using the [*TIMESTEP*] and <<*time*> <*value*>> parameters. The loop command determines the number of steps with the <*n*> parameter, and when *n* = –1, the loop repeats indefinitely. The parameter <*label name*> is an indicator to start the loop, and the name is chosen arbitrarily.

9.2.2   Examples

To illustrate the use of the stimulus generator, consider the following examples.

Example 9.2.2.1
To generate a 0 to 1 transition after 10ns, the following stimulus description statement is required:

```
.LIB C:\DIGITAL.LIB; Reference IO_STM model
*
* Waveform description statement
*
U1 STIM(1,1) $G_DPWR $G_DGND S1 IO_STM 0ns 0 10ns 1
```

where U1 is a user-defined stimulus device, and STIM(1,1) describes one output signal with binary format. The signal and its format are specified in PSpice as (1,1). The signal is described as pairs of time and logic level values. In this case, at 0ns, the logic level is 0, and at 10ns, the logic level is 1. This signal is at node S1, as shown in Figure 9.3.

Example 9.2.2.2
It is necessary to generate the following serial digital word (0010100011000010), assuming a clock step of 2ns. The signal description is given below. This shows that the signal time values have been expressed relative to the previous values because each time value has been prefixed with the "+" sign. For example (0ns 0 +4ns 1 +2ns 0) is equivalent to (0ns 0 4ns 1 6ns 0). Note that the "+" sign at the beginning of some statements shown below indicates continuation of the statement. The digital signal is at node S2, as shown in Figure 9.3.

```
.LIB C:\DIGITAL.LIB; Reference IO_STM model
*
U2 STIM(1,1) $G_DPWR $G_DGND S2 IO_STM
*
+ 0ns 0 +4ns 1 +2ns 0 +2ns 1 +2ns 0 +6ns 1 +4ns 0
+ +8ns 1 +2ns 0; waveform values
```

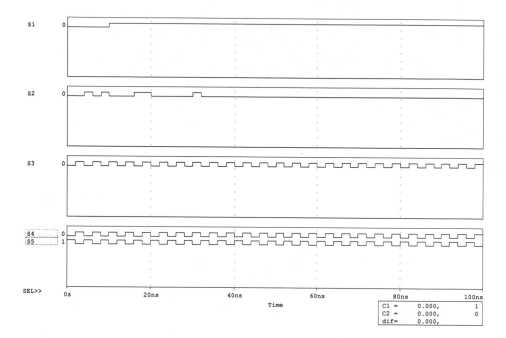

Figure 9.3   Digital waveforms of Examples 9.2.2.1 through 9.2.2.4.

Example 9.2.2.3

This example covers the generation of a clock signal that has a step of 2ns. Repetitive waveforms are best described using the [*TIMESTEP*], and the loop command discussed earlier. The clock signal description is

```
.LIB C:\DIGITAL.LIB; Reference IO_STM model
*
U3 STIM(1,1) $G_DPWR $G_DGND S3 IO_STM TIMESTEP=2ns
*
+ 0C 0           ; at time 0 seconds, the clock is at level 0
*
+ LABEL=LOOP     ; loop label
*
+ 1C 1           ; at time=2ns, the clock is set to 1
*
+ 2C 0           ; at time=4ns, the clock is set to 0
*
+ 3C GOTO LOOP -1 TIMES  ; at time=6ns, branch to LOOP and
*                        ; repeats the loop indefinitely.
```

where C represents a time step, and in this case it is 2ns. The number –1 in the GOTO statement indicates an infinite loop. The label "LOOP" has been chosen arbitrarily. The clock signal is generated at node S3, as shown in Figure 9.3.

Example 9.2.2.4
Often the generation of a nonoverlapping two-phase clock signal is required. These signals are described similar to that of Example 9.2.2.3, but in this case the STIM parameter will have the values (2,11), which means two signals (2) are required, each of binary format (11). The description of the signals is as follows:

```
.LIB C:\DIGITAL.LIB; Reference stimulus generator I/O model
*
U4 STIM(2,11) $G_DPWR $G_DGND S4 S5 IO_STM TIMESTEP=2ns
*
+ 0C 01       ; at time=0ns, 1st signal is set to level 0
*             at time=0ns, 2nd signal is set to level 1
+ LABEL=LOOP
+ 1C 10       ; at time=2ns, 1st signal is set to level 1
+             at time=2ns, 2nd signal is set to level 0
*
+ 2C 01       ; at time=4ns, 1st signal is set to level 0
+             ; at time=4ns, 2nd signal is set to level 1
*
+ 3C GOTO LOOP -1 TIMES  ; at time=6ns, goto LOOP and
*                        ; repeats the loop indefinitely.
```

The two clock signals are generated at nodes S4 and S5, respectively, as shown in Figure 9.3. Note that all the digital waveforms shown in Figure 9.3 were produced by *Probe* in conjunction with a .TRAN statement of the following form:

.TRAN .1ns 100ns

9.3   Simulation examples

Having shown how digital components are described and how digital waveforms are generated, we may now use PSpice to obtain the response of digital circuits. Two examples are considered: the first is based on simple gates, while the second is based on a combination of gates and edge-triggered flip-flops.

Example 9.3.1
The standard gate circuit is shown in Figure 9.1. It consists of a two-input NAND gate and a two-input NOR gate. The PSpice input file of the circuit is given in Listing 9.1. This shows that gates are described in terms of input nodes, output node, timing, and input/output models. Figure 9.4 shows the PSPice input and output waveforms of the circuit.

Example 9.3.2
Consider the circuit in Figure 9.5. It consists of standard logic gates and a positive-edge-triggered D-type flip-flop.

Listing 9.1   PSpice Input File of Example 9.3.1

```
Circuit Description (Figure 9.1)
*
.LIB C:\DIGITAL.LIB; Reference IO_STM model
*
* Digital component description
*
U1 NAND(2) $D_DPWR $D_DGND A B X T1 IO1; 2 input NAND gate
*
U2 NOR(2) $D_DPWR $D_DGND C X Y T1 IO1 ; 2 input NOR gate
*
.MODEL T1 UGATE ; Default timing model of both gates
.MODEL IO1 UIO   ; Default I/O model of both gates
*
* Digital Waveform generation
*
Ua STIM(1,1) $D_DPWR $D_DGND A IO_STM
+ 0s 1 4us 0 6us 1        ; A waveform description
*
Ub STIM(1,1) $D_DPWR $D_DGND B IO_STM
+ 0s 0 4.5us 1 12us 0    ; B waveform description
*
Uc STIM(1,1) $D_DPWR $D_DGND C IO_STM
+ 0s 0 7us 1 9us 0        ; C waveform description
*
.TRAN 0.1us 15us          ; transient analysis range
*
.PROBE
*
.END
```

The PSpice description of the circuit is given in Listing 9.2. Note that PSpice defaults the output of flip-flops and latches to the X state (undefined logic state) at the beginning of the simulation. The X state, however, can be overridden by setting the .OPTIONS DIGINITSTATE to either 0 or 1. If it is set to 0, all flip-flops and latches in the circuit will be cleared; if it is set to 1, all flip-flops and latches will be preset. Listing 9.2 shows that the D-type flip-flop of the circuit (Figure 9.5) has been cleared. Also, the preset and clear nodes of this flip-flop have been set to 1 or high. This is achieved using the "Uhi" description statement (see listing), where (2,11) indicates two signals (2), each of binary format (11). The preset and clear signals are described by 0s 11, which means that at time = 0s, the preset (PRE) signal is set to level 1 and the clear (CLR) signal is set to level 1 also. The simulated input and output waveforms of the circuit are shown in Figure 9.6. Note that the timing and I/O model parameters of the gates in the circuit (Figure 9.5) have been assumed to have their default values.

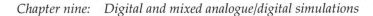

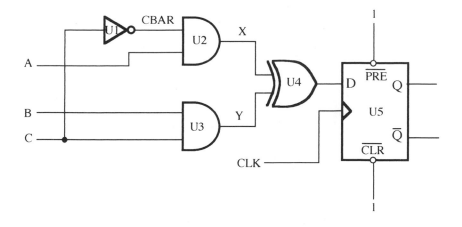

Figure 9.4   Input and output waveforms of the circuit shown in Figure 9.1.

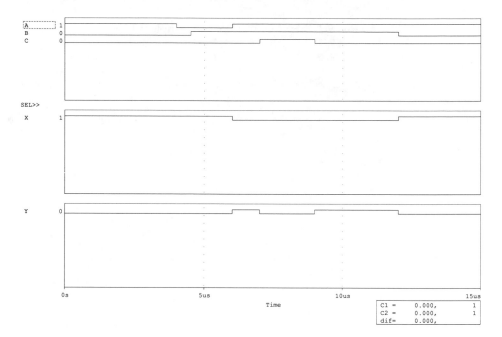

Figure 9.5   Circuit of Example 9.3.2.

9.4   PSpice digital components library

Thus far, we have seen that three statements, digital primitive, timing, and I/O model, are required to describe a digital component. To simplify this process, PSpice offers libraries of digital components and devices, where each component or device is described by a subcircuit. Table 9.6 lists the contents of the PSpice digital library.

Listing 9.2   PSpice Input File of Example 9.3.2

```
Circuit Description (Figure 9.5)
*
.LIB C:\DIGITAL.LIB; Reference IO_STM model
*
.OPTIONS DIGINITSTATE=0; Clearing flip-flop output
*
* Digital components description
*
U1 INV $D_DPWR $D_DGND C CBAR T1 IO1        ; inverter gate
U2 AND(2) $D_DPWR $D_DGND A CBAR X T1 IO1 ; 2 input AND gate
U3 AND(2) $D_DPWR $D_DGND B C Y T1 IO1    ; 2 input AND gate
U4 XOR(2) $D_DPWR $D_DGND X Y D T1 IO1    ; 2 input XOR gate
*
* D-type FF description
*
U5 DFF(1) $D_DPWR $D_DGND PRE CLR CLK D Q QBAR T2 IO1
*
* Default gates and FF timing and I/O models
*
.MODEL T1 UGATE   ; default timing model of standard gates
.MODEL T2 UEFF    ; default timing model of edge-triggered FF
.MODEL IO1 UIO    ; default I/O model of all digital
components
*
* Digital waveforms generation
*
Ua STIM(1,1) $D_DPWR $D_DGND A IO_STM
+ 0s 0 4us 1 18us 0   ; A waveform description
*
Ub STIM(1,1) $D_DPWR $D_DGND B IO_STM
+ 0s 0 5us 1 13us 0   ; B waveform description
*
Uc STIM(1,1) $D_DPWR $D_DGND C IO_STM
+ 0s 0 2us 1          ; C waveform description
*
Uclk STIM(1,1) $D_DPWR $D_DGND CLK IO_STM TIMESTEP=1us
+ 0C 0
+ LABEL=LOOP
+ 1C 1
+ 2C 0
+ 3C GOTO LOOP -1 TIMES  ; Clock waveform description
*
*
Uhi STIM(2,11) $D_DPWR $D_DGND PRE CLR IO_STM
+ 0s 11                  ; FF pre and clr waveforms
*
.TRAN 0.1us 30us         ; transient analysis range
*
.PROBE
*
.END
```

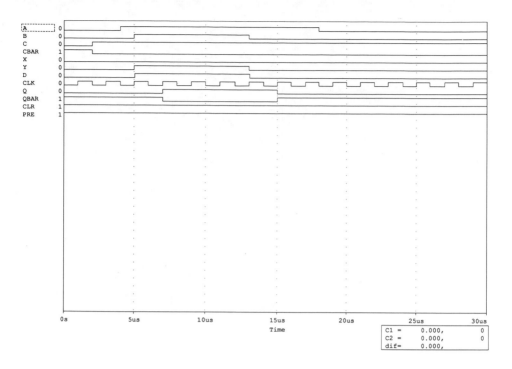

Figure 9.6   Input and output waveforms of Example 9.3.2.

Table 9.6   PSpice Digital Library

File name	Contents
dig_1.lib	74XX00-74XX159
dig_2.lib	74XX160-74XX280
dig_3.lib	74XX281-74XX649
dig_4.lib	74XX650-74XX29828
dig_5.lib	CD4000
dig_ECL.lib	10K and 1K ECL devices
dig_PAL.lib	PAL devices
dig_GAL.lib	GAL devices
dig_misc.lib	Pull-up/-down resistors, delay line
dig_IO.lib	I/O models, AtoD, DtoA interface subcircuits
digital.lib	Master library, references each of the above libraries

This shows, for example, that device models of the 74 TTL series
and the 4000 CMOS series are available, where the parameter "xx"
indicates the device technology type. Table 9.7 shows, for example, how
a 7400 device (Quad, 2-input NAND) and 7474 device (Dual D-type
flip-flop) would be listed in the PSpice library. The sign "*" means the
device model is available in this technology, "A" means the device is

Table 9.7 Extract From PSpice Digital Library

Type	Meaning	TTL	AC	ACT	ALS	AS	F	H	HC	HCT	L	LS	S	Pin list
00	2-Input NAND gate	*	*	*	A	*	*	*	*		54	*	*	A B Y
74	Flip-flop D-type with preset and clear	*			A	*	*	*	*	*	*	A	*	1CLRBAR 1D 1CLK 1PREBAR 1Q 1QBAR

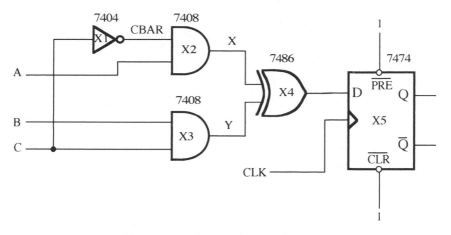

Figure 9.7 Circuit of Example 9.4.1.

called "74ALS00" to differentiate it from 74LS00, and "54" means that the device model is called "54L00". In the digital library, subcircuit names for digital devices are always the commercial part names. The nodes of the subcircuit are given in the "Pin list" column, and the order of the subcircuit nodes is usually "input" followed by "output" nodes. For example, in the case of the 7400 device, the inputs are represented by the nodes A and B, while the output is represented by the node Y.

PSpice has more than 1800 digital component models in the library. A complete list of the digital device models is given in the PSpice manual[1].

To describe a particular digital component, all that is required is an X call subcircuit statement (see Section 5.2, Chapter 5). To illustrate this, consider the following example.

Example 9.4.1
Figure 9.7 shows a circuit, which has already been described using digital primitive description statements as shown in Listing 9.2 and Figure 9.5.

However, using the digital library, the circuit description is given in Listing 9.3. The PSpice description of library digital components basically involves an X call statement, followed by the input(s) and output(s) nodes and the name of the selected subcircuit. For example, the inverter gate has been described by X1, followed by input (C), output (CBAR), and the subcircuit name (7404). Similarly, the exclusive OR gate has been described by X2, followed by inputs (X and Y), output (D), and the subcircuit name (7486). Although the actual 7408 chip contains four 2-input AND gates, the two AND gates in circuit (Figure 9.7) have been described using two 7408 subcircuits. This is because

Listing 9.3   PSpice Input File of Example 9.4.1

```
Circuit Description (Figure 9.7)
*
.LIB C:\DIGITAL.LIB          ; Reference digital.library
.OPTIONS DIGINITSTATE=0       ; Clear flip-flops
*
* Digital components
*
X1 C CBAR 7404       ; Inverter description
X2 A CBAR X 740      ; 2 input AND gate description
X3 B C Y 7408        ; 2 input AND gate description
X4 X Y D 7486        ; 2 input Exclusive OR gate description
X5 CLR D CLK PRE Q QBAR 7474 ; D_type flip-flop description
*
* Waveforms generation
Ua STIM(1,1) $D_DPWR $D_DGND A IO_STM
+ 0s 0 4us 1 18us 0            ; Waveform A description
*
Ub STIM(1,1) $D_DPWR $D_DGND B IO_STM
+ 0s 0 5us 1 13us 0            ; Waveform B description
*
Uc STIM(1,1) $D_DPWR $D_DGND C IO_STM
+ 0s 0 2us 1                   ; Waveform C description
*
*
Uclk STIM(1,1) $D_DPWR $D_DGND CLK IO_STM TIMESTEP=1us
+ 0C 0
+ LABEL=LOOP
+ 1C 1
+ 2C 0
+ 3C GOTO LOOP -1 TIMES      ; Clock waveform description
*
*
Uhi STIM(2,11) $D_DPWR $D_DGND PRE CLR IO_STM
+ 0s 11                      ; FF pre and clr signals
*
* Flip-flop (Qbar) initially set to high impedance
*
Uz STIM(1,1) $D_DPWR $D_DGND QBAR IO_STM
+ 0s Z
*
.TRAN 0.1us 30us             ; Transient analysis range
*
.PROBE
*
.END
```

there is only one AND gate in the 7408 subcircuit. This is true for all the chips that contain multiple identical circuits. The D-type flip-flop has been described using the 7474 subcircuit. Note that the D-type flip-flop nodes have been specified according to the pin list of Table 9.7. It has been assumed that all the gates and the flip-flop of the circuit shown in Figure 9.7 are of the standard TTL type.

This example shows that it is easier to describe digital components in terms of subcircuits from the digital library than using digital primitive description, timing, and I/O model statements (Listing 9.2).

9.5 Modifying digital model parameters

PSpice digital component subcircuits contain timing and I/O model information. PSpice allows the user to describe the timing and I/O model values in terms of minimum, typical, or maximum values. Each subcircuit has two optional parameters: MNTYMXDLY and IO_LEVEL. The first parameter (MNTYMXDLY) allows the user to specify minimum, nominal, or maximum values for the component timing characteristics such as propagation delay, setup, and hold times. This is achieved by assigning digits 1, 2, or 3, respectively, to the parameter MNTYMXDLY. If MNTYMXDLY is not specified by the user, PSpice will set the propagation delay to nominal values (i.e., PSpice defaults to digit 2). In Listing 9.3, the gates and the flip-flop were assumed to have nominal timing model characteristics (since no value for the parameter MNTYMXDLY was specified in the subcircuit descriptions). Box 9.1 shows the nominal timing model values used for the digital components of circuit (Figure 9.7 and Listing 9.3). These values are obtained from the PSpice circuit output file. The column D_04, for example, gives the inverter gate (7404) timing model nominal parameter values. These parameters are defined in Table 9.2.

The second subcircuit parameter (IO_LEVEL) allows the user to specify the complexity of the I/O model of the digital component. There are four models available, which are chosen by specifying the digits 1,2, 3, and 4, respectively, to the parameter IO_LEVEL. PSpice defaults to level 1, which is sufficient for simple circuit simulations. The higher levels should be used only when a high degree of simulation accuracy is required. For example, if the propagation delay of the inverter gate (Figure 9.7), which is described by X1 (Listing 9.3), is required to have maximum delay, instead of the default typical delay, the inverter gate description must be modified to

X1 C CBAR 7404 PARAMS: MNTYMXDLY=3, IO_LEVEL=2

This statement also assumes a more complex and accurate I/O inverter model.

Box 9.1 Timing Model Values of the Circuit (Figure 9.7)

```
**** Digital Gate MODEL PARAMETERS
****************************************************************

                D_04              D_08              D_86

TPLHMN      4.800000E-09       7.000000E-09      3.600000E-09
TPLHTY     12.000000E-09      17.500000E-09      9.000000E-09
TPLHMX     22.000000E-09      27.000000E-09     17.000000E-09
TPHLMN      3.200000E-09       4.800000E-09      2.000000E-09
TPHLTY      8.000000E-09      12.000000E-09      5.000000E-09
TPHLMX     15.000000E-09      19.000000E-09     11.000000E-09


**** Digital Edge-Triggered FF MODEL PARAMETERS
****************************************************************

                   D_74
TPCLKQLHMN        5.600000E-09
TPCLKQLHTY       14.000000E-09
TPCLKQLHMX       25.000000E-09
TPCLKQHLMN        8.000000E-09
TPCLKQHLTY       20.000000E-09
TPCLKQHLMX       40.000000E-09
TPPCQLHMN         6.250000E-09
TPPCQLHTY        15.625000E-09
TPPCQLHMX        25.000000E-09
TPPCQHLMN        10.000000E-09
TPPCQHLTY        25.000000E-09
TPPCQHLMX        40.000000E-09
TWCLKLMN         37.000000E-09
TWCLKLTY         37.000000E-09
TWCLKLMX         37.000000E-09
TWCLKHMN         30.000000E-09
TWCLKHTY         30.000000E-09
TWCLKHMX         30.000000E-09
TWPCLMN          30.000000E-09
TWPCLTY          30.000000E-09
TWPCLMX          30.000000E-09
TSUDCLKMN        20.000000E-09
TSUDCLKTY        20.000000E-09
TSUDCLKMX        20.000000E-09
TSUPCCLKHMN      0
TSUPCCLKHTY      0
TSUPCCLKHMX      0
THDCLKMN          5.000000E-09
THDCLKTY          5.000000E-09
THDCLKMX          5.000000E-09
```

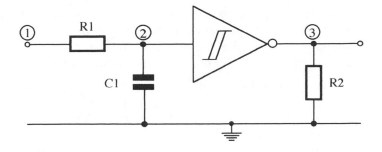

Figure 9.8   Simple mixed analogue/digital circuit.

9.6   Mixed analogue/digital circuit simulations

PSpice recognises three types of nodes: analogue, digital, and interface. The type of node is determined by the types of components connected to it. If all the components connected to a node are analogue, then the node is analogue. Similarly, if all the components connected to a node are digital, then the node is digital. If there is a combination of analogue and digital components connected to a node, then the node is interface. Such nodes occur in mixed analogue/digital circuits, as shown in Figure 9.8, where node 1 is a pure analogue node and nodes 2 and 3 are interface nodes.

PSpice automatically breaks interface nodes into one purely analogue or digital node using analogue/digital or digital/analogue interface subcircuits. These interface subcircuits are described in PSpice as AtoD and DtoA, respectively. The function of an AtoD interface subcircuit is to change voltages and impedances to digital states (Table 9.5). Similarly, the function of a DtoA interface subcircuit is to change digital states to voltages and impedances. Figure 9.9 shows the circuit

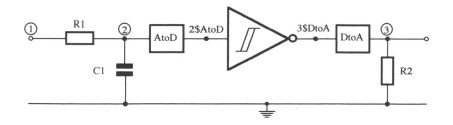

Figure 9.9   Circuit of Figure 9.8 after PSpice has inserted AtoD and DtoA interface subcircuits.

of Figure 9.8 after PSpice has inserted AtoD and DtoA interface subcircuits. Nodes 2 and 3 are now purely analogue nodes. PSpice has created two new digital nodes designated 2$AtoD and 3$DtoA.

PSpice uses the following notation when creating those new nodes:

1. The analogue nodes retain the name of the original interface nodes as shown in Figure 9.9 with nodes 2 and 3.
2. Each new digital node name is suffixed with "$AtoD" or "$DtoA". For example in the circuit shown in Figure 9.9, the digital nodes are 2$AtoD, and 3$DtoA. If the analogue node is attached to more than one digital device, the second digital node is appended with "$AtoD2" or "$DtoA2", and so on.

The AtoD and DtoA interface subcircuits are part of the I/O model of the digital device (see Table 9.6). The user does not need to generate interface subcircuits; all that is required is to specify the digital library in the circuit-input file using the statement

.LIB C:\digital.lib

This assumes that the digital library is on the C drive of the PC. The use of PSpice in performing mixed analogue/digital simulation is demonstrated in the following example.

Example 9.6.1
Consider the simulation of the simple gate oscillator circuit shown in Figure 9.10. It consists of a Schmitt trigger, a basic inverter, and an RC network.

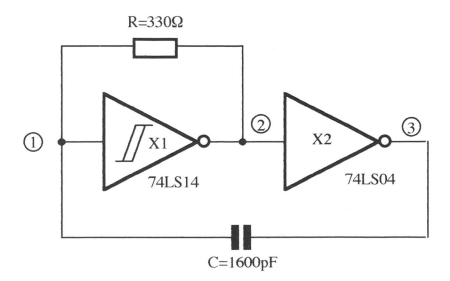

Figure 9.10 Circuit of Example 9.6.1.

Listing 9.4 PSpice Input File of Example 9.6.1

```
Gates based oscillator circuit (Figure 9.10)
*
.LIB C:\DIGITAL.LIB          ; digital components library
*
* Digital components
X1 1 2 74LS14                ; schmitt trigger inverter
X2 2 3 74LS04                ; basic inverter
*
* Passive component
R 1 2 330
C 1 3 1600pF
*
.TRAN 1ns 5us               ; Transient analysis range
*
.PROBE
*
.END
```

The PSpice description of the oscillator circuit is given in Listing 9.4. The analogue and digital waveforms of the oscillator are given in Figure 9.11. Note that *Probe* recognises the new digital nodes PSpice introduces when changing interface nodes to purely analogue or digital.

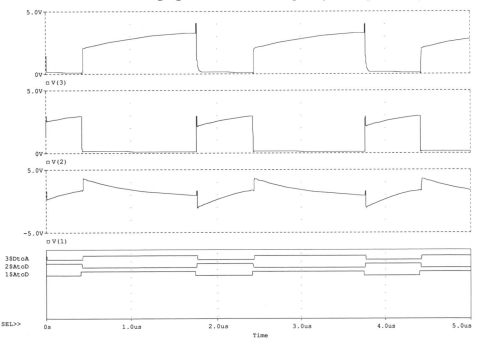

Figure 9.11   Analogue and digital waveforms of Example 9.6.1.

9.7   Chapter summary

This chapter has only briefly introduced the subject of digital and mixed analogue/digital simulation with PSpice. To simplify the process of digital component description, PSpice has an extensive library containing models of commonly used digital components, including the 74 TTL series and the 4000 CMOS series. These models are described in the form of subcircuits, which are readily called using X statements. A full description of the digital simulator is beyond the scope of this book, but the power of this simulator must not be underestimated. It should be noted that the simulation of complex digital circuits will require the use of the schematic capture (see Chapter 2, Section 2.8).

Reference

1. MicroSim Corporation, *Circuit Analysis — Reference Manual* (The Design Center), Version 5.3, Irvine, CA, January 1993.

appendix A

PSpice error messages

This appendix lists some of the common error messages reported by PSpice when there is an error in the circuit input file. The error messages usually appear in the circuit output file.

A.1 Nodes with LESS than two connections

PSpice requires that each and every node of a circuit must be connected to at least two other nodes. Otherwise PSpice will report the following error message:

Error — Less than two connections at node x

where x is an arbitrarily chosen node. The usual way around this problem is to connect a component that has no effect on the circuit. A typical situation is shown in Figure A.1, where node 3 has only one connection, and in this case a large-value resistor (1GΩ, say) is connected from node 3 to ground.

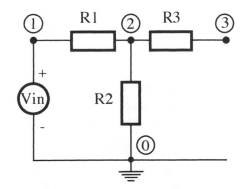

Figure A.1 Typical circuit with less than two connections at node 3.

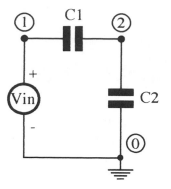

Figure A.2 Typical circuit without DC path.

A.2 Floating nodes

PSpice requires that every node has a DC path to ground, that is, there should be no "floating nodes". If there are any, PSpice will report the following error message:

> Error — Node x is floating

This means that there is no DC path from node x to ground. From the PSpice point of view, a DC path allows current to flow to ground, which may be via resistors, inductors, and semiconductor components. If a circuit defines a node that is floating (e.g., node 2 of the circuit in Figure A.2), then it is necessary to connect a component that provides the DC path, but has insignificant effect on the circuit operation. In the case of the circuit in Figure A.2, a resistor of 1GΩ connected from node 2 to ground will provide the DC path and not affect the circuit.

A.3 Voltage source and inductor loops

PSpice requires that there should be no loops with zero resistance (i.e., short circuit loops). Otherwise PSpice will report an error message of the form

> Error — Voltage loop involving <........>

where <.......> gives some additional information regarding the error. The zero-resistance components in PSpice are independent voltage sources (V), inductors (L), voltage-controlled voltage sources (E), and current-controlled voltage sources (H). A typical circuit with such a loop is shown in Figure A.3. PSpice will report the following error message:

> Error — Voltage loop involving Vin

The solution to this problem is to add a small series resistance (say, 0.001Ω) to at least one component in the loop. The value of this resistor is chosen so that it has no effect on the circuit operation.

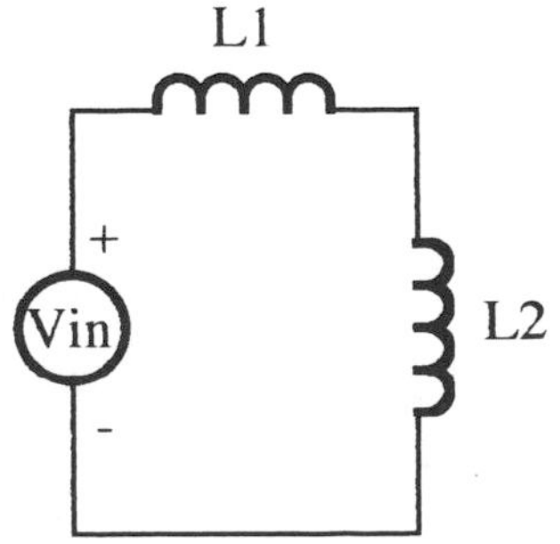

Figure A.3 Typical circuit with zero-resistance loop.

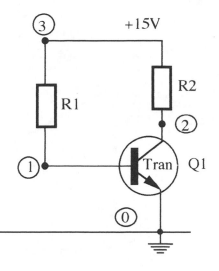

Figure A.4 Typical circuit illustrating the error message discussed in Section A.4.

A.4 Missing .MODEL statements

User-defined semiconductor components are described in PSpice using two statements, a component and a .MODEL statement. If the .MODEL statement is missing, PSpice will report an error message of the form

Error — Model <...> used by <...> is undefined

As an example, if the .MODEL statement of the transistor Q1 (Figure A.4) is omitted from the input file, PSpice will report the following error message:

Error — Model Tran used by Q1 is undefined

Clearly, the solution is to include the .MODEL statement (see Chapter 2, Example 2.7.2).

A.5 Unspecified component model libraries

When a library model for a particular component is selected, the library name of the component model must be specified. Otherwise PSpice will report error messages of the form

Error — Model <....> used by <....> is undefined
Error — Subcircuit <....> used by <....> is undefined

As an example, if the diode and the op-amp model libraries of the circuit in Figure A.5 are not specified in the input file, PSpice will report

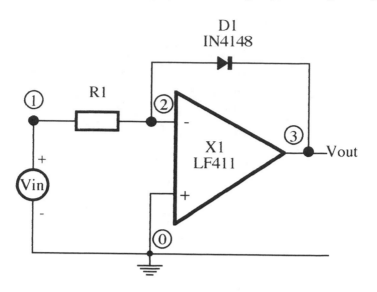

Figure A.5   Typical circuit illustrating the error message discussed in Section A.5.

the following error messages:

  Error — Model D1N4148 used by D1 is undefined
  Error — Subcircuit LF411 used by X1 is undefined

Clearly, the solution to this problem is to specify the appropriate model libraries using the .LIB statement. In the case of the diode, a statement of the following form is needed:

.LIB C:\diode.lib

and in the case of the op-amp, a statement of the following form is required:

.LIB C:\linear.lib

Both statements assume that the PSpice diode and op-amp model libraries are on drive C of the PC. Note that in the case of the op-amp model, the linear library has been selected. Other op-amp model libraries are also available (see Table 6.9 of Chapter 6).

A.6   Unable to finish transient analysis

During transient analysis, PSpice occasionally reports the following error message on the screen: "Unable to finish transient analysis". Commonly (but not always), this problem might be solved if the parameter RELTOL (see Appendix F) is relaxed from its default of 0.001 to

0.01, say. The value of RELTOL is inversely proportional to simulation time and memory space. To set the relaxed value of RELTOL, the following .OPTION statement is required:

.OPTIONS RELTOL=0.01

A.7 Out of memory

PSpice occasionally reports the following error message on the screen: "Out of memory". This problem might be solved if the parameter RELTOL is relaxed, as discussed in Section A.6, or the random access memory in the machine is increased.

appendix B

Stimulus Editor (StmEd)

PSpice has a number of time-dependent sources that can be used to provide circuit input signals. These sources can produce voltage or current signals that have the following description: sinusoidal (SIN), exponential (EXP), pulse (PULSE), single-frequency frequency-modulated (SFFM), and piece-wise linear (PWL). The PWL source may be used to approximate any given signal or waveform. Each of these signals is described in terms of a number of parameters as discussed in Chapter 3 (Tables 3.1 to 3.5). Entering these parameters for a particular signal in the right format and order can easily prove to be a time-consuming and error-prone task. To simplify and partially automate the process of generating waveforms, PSpice offers the user the program *Stimulus Editor (StmED)*. This program allows the user to set up and verify the input signal(s) for a simulation. Menu prompts guide the user to provide the necessary signal parameters, such as rise time, fall time, signal period, or the (value, time) coordinates of the waveform, as described using the PWL source. Graphical feedback allows the user to verify the waveform quickly.

The *StmEd* program is activated by selecting the option "StmEd" from the *Control Shell* program (see Chapter 2, Section 2.6). The *StmEd* program requires the presence of a circuit input file.

Example B.1

A specific application requires the train of data (1000110111) to be used as an input to some analogue circuitry, where "1" represents a 5V amplitude and 144ns pulse period, and "0" represents a 0V and 144ns pulse period. This signal is best described in PSpice using PWL voltage source. Assume that the input file of the circuit to be simulated has been created using the editor of the *Control Shell* program. To generate the input signal, select the option "StmEd" from the *Control Shell* program. The program will ask the user the following:

1. Type of source (voltage or current)
2. Type of signal (SIN, EXP, PULSE, SFFM, or PWL)
3. Signal parameters

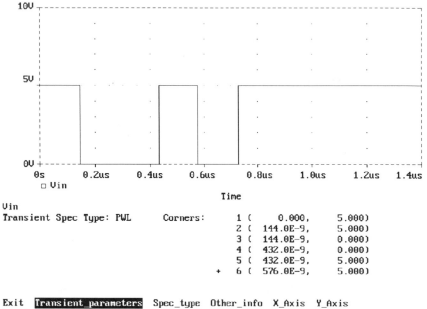

Figure B.1   The 1000110111 waveform generated by *StmEd*.

In this case, voltage and PWL will be chosen, and then 10 points of (time, voltage) will be typed. This number of points describes the required input data. The program will then plot the waveform shown in Figure B.1. Note that *StmEd* requires the time value coordinates to increase from one value to the next. Hence, when describing an "instantaneous" signal value the same time value cannot be used; therefore, a slightly higher time value must be specified.

The program *StmEd* has editing facilities that allows the user to modify or change the entered data. If the signal is OK, then the program will automatically add it as an input to the circuit input file. All that is required now is to specify the nodes between which the signal source is connected.

If you have the digital simulation option of PSpice, digital waveforms can be created using the stimulus generator (Chapter 9, Section 9.2.1). However, it is not possible to specify logic level "1", for example, to have a specific voltage value. This is because logic level "1" in the digital simulator only means that the voltage is somewhere within the "high" range of a particular digital device.

appendix C

Simulation of oscillator circuits

This appendix shows how PSpice can be used to simulate oscillators. An oscillator is a device that produces a periodic AC output signal without any form of specified input signal. A widely used low-frequency oscillator is shown in Figure C.1. This circuit is the Wein bridge oscillator.

For the Wein bridge circuit to oscillate, the following condition must be met:

$$A\beta = 1\angle 0°$$

where A is the amplifier closed loop gain and is given by $(1+RB/RA)$, and β is the feedback voltage fraction and is given by

$$\beta = \frac{Z1}{Z1+Z2}$$

where $Z1$ is

$$Z1 = \frac{R1}{sR1C1+1}$$

and $Z2$ is

$$Z2 = \frac{sR2C2+1}{sC2}$$

where s is the Laplace transform variable, that is, $s=j\omega$. Substituting for $Z1$ and $Z2$ in the expression for β gives

$$\beta = \frac{sR1C2}{\left(s^2R1R2C1C2 + s(R1C1 + R2C2 + R1C2) + 1\right)}$$

It can been shown[1] that at one particular frequency the oscillation condition is met. This frequency is known as the frequency of oscillation, and it occurs when $A=3$, $\beta=1/3$ and has zero phase shift. The frequency of oscillation, F, is given by

$$F = \frac{1}{2\pi RC}$$

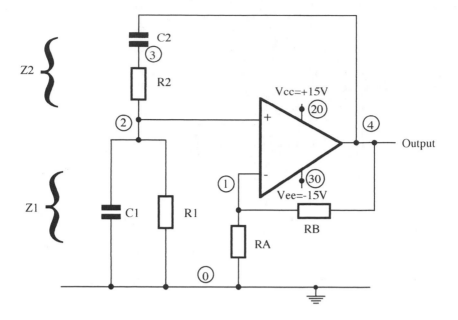

Figure C.1   Wein bridge oscillator.

assuming that R1=R2=R and C1=C2=C.

The simulation of oscillator circuits using PSpice usually produces no output or the error message "Unable to calculate transient analysis bias points". This is because the oscillator theoretically has no input signal. In practice, an oscillator starts oscillating because there is a small voltage or noise on the noninverting input or the output of the amplifier. This small voltage must be included in the simulation of oscillator circuits to overcome these start-up problems. To illustrate this point, consider the following example.

Example C.1

Simulate the response of the Wein bridge oscillator shown in Figure C.1 with the following component values:

RA = 10kΩ, RB = 20kΩ, R1 = R2 = 6.366kΩ, C1 = C2 = 1nF

To start oscillation, a small voltage at the amplifier output is required. This is achieved by using the .IC (Initial Condition) statement as shown in Listing C.1. This shows that V(4) has been set arbitrarily to 0.1V. The voltage specified in the .IC statement can be considered as an input signal to the oscillator.

The amplifier has been assumed to be the "LF411", the model of which has been taken from the "LINEAR.LIB" PSpice library (see Chapter

Listing C.1   PSpice Input File of Circuit (Figure C.1)

```
Wein Bridge Oscillator
*
.LIB  C:\LINEAR.LIB               ; op-amp model library
.TRAN 0.1u 700u 0u 0.1u
*
Vcc 20 0 15V                      ; 15V power supply
Vee 30 0 -15V                     ; -15V power supply
**
RA 1 0 10k
RB 1 4 20K
R1 2 0 6.366k
C1 2 0 1nF
R2 2 3 6.366k
C2 3 4 1nF
**
Xop-amp 2 1 20 30 4 LF411         ; op-amp description
.IC V(4)=0.1V                     ; set initial condition
*
.PROBE V(4)
.END
```

6 Section 6.5). Figure C.2a gives the simulated oscillator output, which shows that the frequency of oscillation is about 25kHz, as expected. Had the initial condition been set to 0.1V at node 2 (the noninverting amplifier input), the output shown in Figure C.2b would have been produced. This simulation shows that different starting conditions produce the correct oscillation frequency but with different amplitudes. Figure C.2 was produced by combining Listing C.1 with a modified Listing C.1, (that is .IC V(2)=0.1V in place of .IC V(4)=0.1V), as one input file, and running one analysis.

As mentioned earlier, the amplifier closed loop gain (*A*) must be 3 in order to produce a sine wave oscillation. Increasing RB to 21kΩ causes the oscillator to saturate, as shown in Figure C.3a. Reducing RB to 19kΩ causes the oscillator to stop oscillating, as shown in Figure C.3b. This illustrates that the amplifier gain condition of a Wein bridge oscillator is very critical and therefore that all practical circuits include some form of amplitude stabilisation as shown in Figure C.4.

The PSpice input file of the modified Wein bridge oscillator is given in Listing C.2. It has been assumed that the diodes are the "1N4148" type, and the model is taken from the "DIODE.LIB" PSpice library. Simulating the modified Wein oscillator (Figure C.4) produces the output shown in Figure C.5, which shows the correct oscillation frequency of about 25kHz.

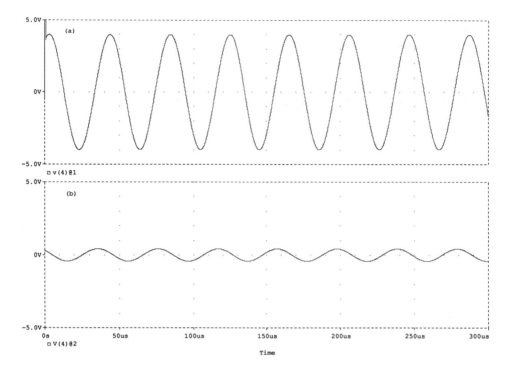

Figure C.2   Wein bridge oscillator response when (a) V(4)=0.1V; (b) V(2)=0.1V.

Figure C.5a shows the oscillator response with .IC V(4)=0.1V, while Figure C.5b shows the response with .IC V(2)=0.1V. Unlike Figure C.2, this simulation shows that different starting conditions with the modified Wein bridge oscillator produce the same amplitude and oscillation frequency.

Another method that can be used to overcome the oscillator start-up simulation problem is to ramp up one of the oscillator power supplies. For example, replacing the positive power supply (Vcc) description statement shown in Listing C.2 with

Vcc 20 0 PULSE (0V 15V 0s 10ns 10ns 60s 120s)

and removing the .IC statement produces the correct oscillation frequency as shown in Figure C.6. This statement shows that the positive power supply goes from 0 to 15V in 10ns and stays there for a fixed time (60s). The (120ns) parameter represents the period of the pulse, and has been chosen arbitrarily. Note that the start-up oscillation (Figure C.6) is different from those obtained when initial conditions were set (Figure C.5).

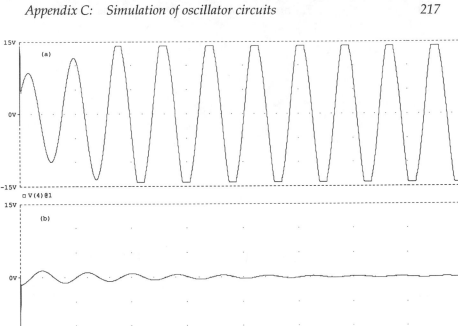

Figure C.3   Wein bridge oscillator response when (a) RB=21kΩ; (b) RA=19kΩ.

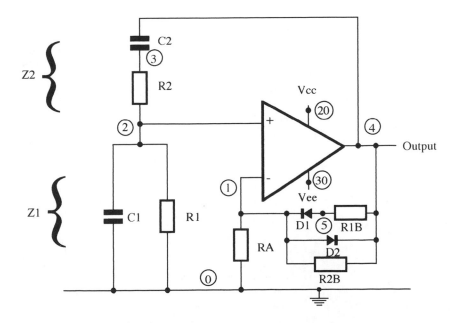

Figure C.4   Modified Wein bridge oscillator including amplitude stabilisation.

Listing C.2   PSpice Input File of Circuit (Figure C.4)

```
Modified Wein Bridge Oscillator
*
.LIB C:\LINEAR.LIB         ; op-amp model library
.LIB C:\DIODE.LIB          ; diode model library
*
.OPTIONS RELTOL=0.01       ; decrease simulation time
.TRAN 0.5us 1.5ms 0us 0.5us
**
Vcc 20 0 15V
vee 30 0 -15V
**
RA 1 0 10K
R1B 5 4 220K
R2B 1 4 22K
R1 2 0 6.366K
C1 2 0 1nF
R2 2 3 6.366K
C2 3 4 1nF
**
Xop-amp 2 1 20 30 4 LF411     ; op-amp description
D1 5 1 D1N4148
D2 1 4 D1N4148                 ; diode description
**
.IC V(4)=0.1V
*
.PROBE V(4)
.END
```

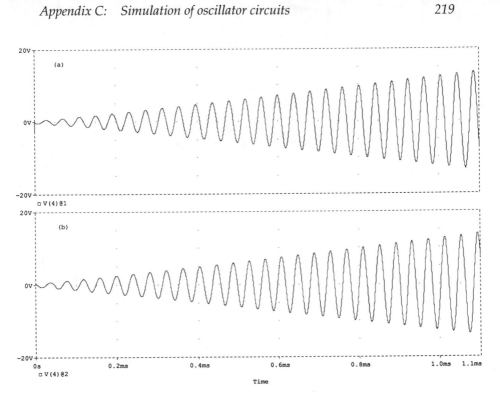

Figure C.5   Modified Wein bridge oscillator response when (a) V(4)=0.1V; (b) V(2)=0.1V.

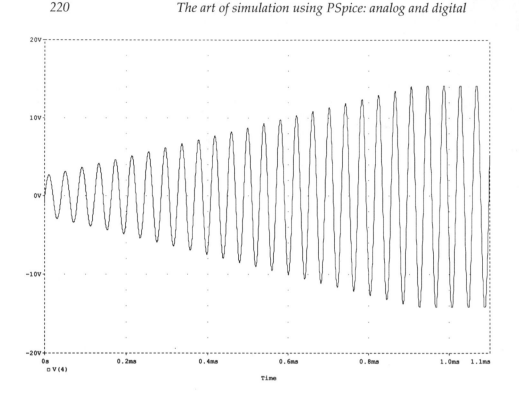

Figure C.6   Modified Wein bridge oscillator response when the positive power supply has been pulsed up.

It is important to remember that an oscillator circuit may take some time to reach the steady-state oscillation. This means that it is worth remembering to adjust the .TRAN statement so that the early stages of the simulation results are suppressed. The conclusion is that PSpice will simulate oscillators provided the initial conditions are set. This may be simply to correctly simulate the way practical power supplies start up and enable oscillation.

Reference

1. Bogart, F.B., *Electronics Devices and Circuits*, 2nd edition, Merrill Publishing Company, Columbus, OH, 1990, pp. 647.

appendix D

Simulation of switched capacitor networks

Switched capacitor (SC) networks work on the principle that a capacitor and a switch can be used to simulate the function of a resistor,[1] as shown in Figure D.1.

The switch is initially in the position shown in Figure D.1a, and the capacitor is charged to the input voltage $V1$. The switch is now thrown (Figure D.1b), and the capacitor is discharged after a determined time to some new voltage, $V2$. The charge transferred is

$$Q = C1(V1 - V2)$$

If the switch is thrown back and forth at a clock rate F_{clk}, the average current flow is given by

$$i = C1(V1 - V2)F_{clk}$$

where F_{clk} is the switching rate or the clock frequency. From Ohm's law, the equivalent resistance of the switched capacitor network is

$$R_{eq} = \frac{V1 - V2}{i} = \frac{1}{C1F_{clk}} \tag{D.1}$$

The switches are normally driven by a nonoverlapping two-phase clock, as shown in the circuit in Figure D.2.

This resistor simulation is attractive from an IC point of view because it is relatively easy to implement accurate capacitors on silicon compared with semiconductor resistors. The IC designer then has the ability to produce full implementation of circuits requiring close-tolerance Rs and Cs such as filters, A/D and D/A converters.

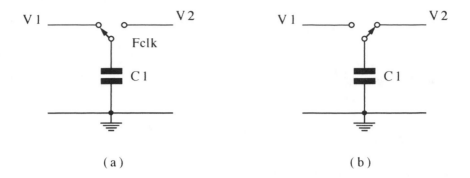

(a)   (b)

Figure D.1 Switched capacitor resistor.

Voltage-controlled switches are recognised in PSpice by the letter S (see Table 2.1 of Chapter 2). Current-controlled switches are also available. Two statements are required to simulate a switch, a description statement and a .MODEL statement. The basic form of the voltage controlled switch is

> *S<name> <+ switch node> <- switch node>*
> *+ <+ controlling node> <– controlling node> <model name>*

where *S* is the PSpice symbol for a voltage-controlled switch and *<name>* is the switch name, which can be up to eight characters long. Examples are S1 and Sclk. The *<+ switch node>* and *<– switch node>* show how the switch is connected within a circuit. The *<+ controlling node>* and *<– controlling node>* define the nodes of the voltage source (clock) that are used to turn the switch on or off. The *<model name>* is usually the descriptive name of the switch.

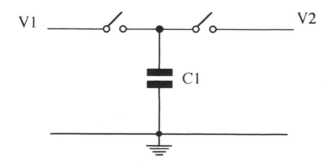

Figure D.2 Switched capacitor resistor with two-phase clock.

Table D.1   Voltage-Controlled Switch Model Parameters

Parameter	Meaning	Default value
RON	"on" resistance	1Ω
ROFF	"off" resistance	$1M\Omega$
VON	Control voltage for "on" state	1V
VOFF	Control voltage for "off" state	0V

The basic form of the .MODEL statement is

    .MODEL <model name> VSWITCH [model parameters]

where *<model name>* is the name given to the switch in the description statement. *VSWITCH* is the PSpice model name of a voltage-controlled switch. PSpice has a model for a voltage-controlled switch, which is described by the parameters shown in Table D.1. These parameters can be specified by the user in the [*model parameters*] part of the .MODEL statement. When the parameters are not specified, PSpice uses the default values. Now it is possible to simulate SC networks using PSpice.

Example D.1
Consider the simple lowpass RC filter shown in Figure D.3, where the filter –3dB cutoff frequency (F_{-3dB}) is given by

$$F_{-3dB} = \frac{1}{2\pi R1C2}$$

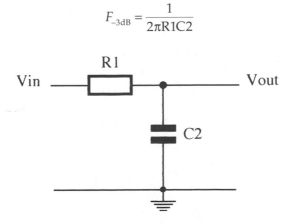

Figure D.3   Simple lowpass RC filter.

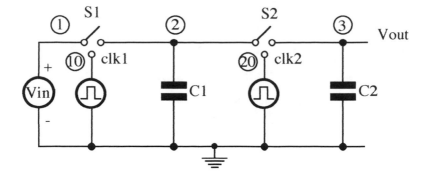

Figure D.4   Equivalent switched capacitor circuit of the filter (Figure D.3).

Substituting Equation (D.1) for R1 gives

$$F_{-3dB} = \frac{F_{clk}C1}{2\pi C2}$$

If F_{-3dB}=10kHz, F_{clk}=500kHz, and C1=20pF, this gives C2=159.15pF.

The equivalent SC network of the RC circuit is shown in Figure D.4, and the input file of the circuit is given in Listing D.1.

Listing D.1   Input File of Example D.1

```
Switched Capacitor Circuit (Figure D.4)
*
.OPTIONS RELTOL=0.01            ; sets simulation accuracy & time
*
Vin 1 0 SIN (0 0.2V 10kHz) ; 10kHz sine wave input signal
*
* Two-phase 500kHz clock
Vclk1 10 0 PULSE (0V 5V 0s 10ns 10ns 1us 2us)
Vclk2 20 0 PULSE (5V 0V 0s 10ns 10ns 1us 2us)
*
.TRAN 0.2us 0.5ms 0.3ms 0.2us ; transient analysis range
*
S1 1 2 10 0 SW1
S2 2 3 20 0 SW1                ; voltage controlled switch
*
* Switch model
.MODEL SW1 VSWITCH (Ron=5 Roff=1e6 Von=5V Voff=0V)
*
C1 2 0 20pF
C2 3 0 159.15pF
*
.PROBE V(1), V(3)             ; graphic output of input & output
*
.END
```

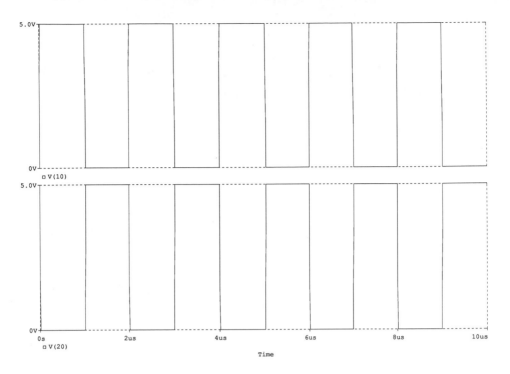

Figure D.5 Nonoverlapping two-phase clock (frequency=500kHz).

It is important to set up PSpice options according to requirements, for example, the option RELTOL (see Appendix F). The default value of RELTOL is 0.001, and changing RELTOL to 0.01 is sufficient in most cases. This change is sometimes necessary for a circuit simulation to complete, as in this example (see Listing D.1). Changing RELTOL to 0.01 reduces computation time and memory space. Failing to do so would result in *Probe*'s reporting the error message "Not enough memory for Fourier transform".

It has been assumed that the input signal is a 10kHz sine wave, with 0.4V pk-pk amplitude. The nonoverlapping two-phase clock (Vclk1 and Vclk2) has been described by a square waveform of 5V amplitude, 10ns rise time and fall time, and 500kHz frequency. The two-phase clock is shown in Figure D.5.

The two switches (S1 and S2) have both been modelled using the model name (SW1). This switch model has an on-resistance of 5Ω, off-resistance of 1MΩ, control voltage for "on" state of 5V, and control voltage for "off" state of 0V. The simulated input and output signals are shown in Figure D.6. The output signal amplitude has been reduced by –3dB (0.707 of the input signal) and has been shifted by –45°. This is because the first-order lowpass filter has a –3dB cutoff frequency of 10kHz.

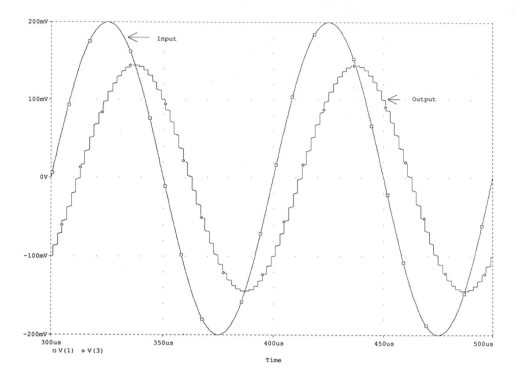

Figure D.6 Input and output signals of the switched capacitor filter (Figure D.4).

The frequency spectrum of the input and the output signals is shown in Figure D.7. Note the modulation components around the switching frequency (500kHz) when a 10kHz input signal is applied to the network. The amplitude of these components has been reduced greatly because of the filter characteristics. The .TRAN statement (Listing D.1) shows that the first 0.3ms of simulation results have been suppressed to reduce the amount of data *Probe* can handle for plotting the signal frequency spectra. Note that the frequency spectra shown in Figure D.7 were obtained as discussed in Section 3.5 of Chapter 3.

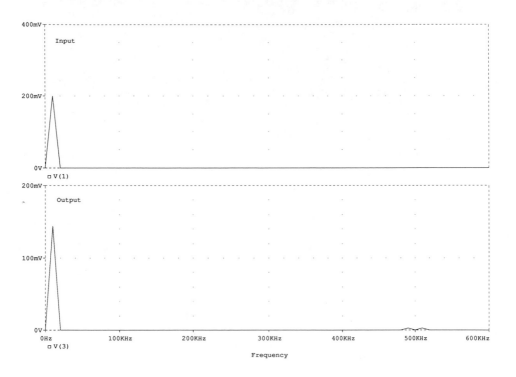

Figure D.7 Frequency spectrum of the input and output signals (Figure D.6).

Reference

1. Van Valkenburg, M.E., *Analogue Filter Design*, Holt, Rinehart and Winston, New York, 1982, pp. 487.

appendix E

Simulation of transmission lines

This appendix covers the simulation of ideal and lossy transmission lines. A transmission line has two ports, A and B, as shown in Figure E.1. The parameter VA is the voltage at port A, and IA is the current into port A. Similarly VB and IB are the voltage and current at port B.

E.1   Ideal transmission lines

An ideal transmission line is described in PSpice using the following description statement:

> *T<name> <A port + node> <A port – node> <B port + node>*
> *+ <B port – node> Z0=<value> [TD=<value>]*
> *+ [F=<value> [NL=<value>]]*

where *T* is the PSpice symbol for a transmission line and *<name>* is the transmission line name, which can be up to eight characters long. The parameters *<A port + node>* and *<A port – node>* are the positive and negative nodes of port A of the transmission line. The parameters for port B are similar. PSpice requires the user to specify the characteristics impedance, *<Z0>*. To complete the ideal transmission line description, either the parameter *<TD>* (transmission delay), or the parameters *<F>*

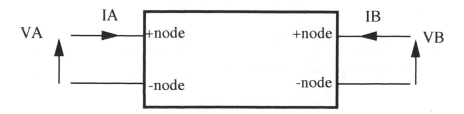

Figure E.1   Block-diagram representation of a transmission line.

(frequency) and <NL> (normalised line length) are required. It should be noted that the transmission delay, TD, is related to line length by the equation

$$TD = (\text{line length/velocity})$$

where velocity is

$$\text{Velocity} = (\text{free space light velocity})*(\text{velocity factor})$$

and

$$\text{Free space light velocity} = 3 \times 10^8 \text{ m/s} \quad \text{Velocity factor} \leq 1$$

The normalised line length, <NL>, is related to the line length by the equation

$$NL = (\text{line length/wavelength})$$

or

$$NL = (\text{line length})*(\text{frequency})/(\text{velocity})$$

where velocity is given by the preceding equation. If <NL> is not specified, PSpice assumes the default value of 0.25, which means the line is to be a one quarter wavelength. To illustrate the use of the above equations, consider the description of a 30m coaxial cable with characteristic impedance of 75Ω, and velocity factor of 0.7.

This transmission line can be described using two methods. First, consider the transmission delay, TD, which is given by

$$TD = 30/(3E8*0.7) = 143\text{ns}$$

so the transmission line description statement is

$$T1\ 2\ 0\ 3\ 0\ Z0 = 75\ TD = 143\text{ns}$$

assuming that 2 and 0 are port A nodes, and 3 and 0 are port B nodes. Second, consider specifying the normalised length, NL, and frequency, F, and assuming F = 20MHz, we have

$$NL = (30*20E6)/(3E8*0.7) = 2.86$$

so the transmission line description statement is

$$\text{T1 2 0 3 0 Z0} = 75 \text{ F} = 20\text{meg NL} = 2.86$$

As far as PSpice is concerned, the two transmission line statements are equivalent. PSpice also allows the parameters *Z0* and *TD* to be specified as expressions. As an example, the second transmission line statement can be specified as

$$\text{T1 2 0 3 0 Z0=\{sqrt(0.378uH/69pF)\} F=20meg NL=2.86}$$

This statement tells PSpice that the characteristic impedance of the line is $\sqrt{\dfrac{L}{C}}$, and in this case the values correspond to coaxial cable type

Belden 8281. The expression must be enclosed within brackets of the type {}. PSpice expressions are discussed in Chapter 5, Section 5.1.

E.2 Lossy transmission lines

Electrical transmission lines consist basically of two conductors separated by an insulator. The conductors have equivalent series resistance and inductance, and the insulator has equivalent shunt conductance and capacitance. Therefore, the equivalent circuit of a transmission line is shown in Figure E.2.

In the previous section it is assumed that the conductors and insulators are perfect. In reality, the conductors and insulators possess resistance, inductance, conductance, and capacitance. As a result, a practical transmission line exhibits attenuation and dispersion. Attenu-

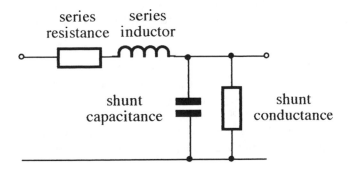

Figure E.2 Equivalent circuit of a transmission line.

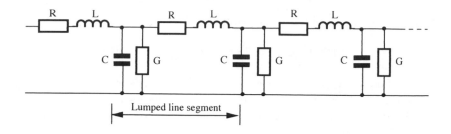

Figure E.3   Lossy transmission line based on the lumped approach.

ation causes signal power loss and is due to the resistance of the conductor and the conductance of the insulator. Note that the signal power loss increases with frequency, which implies that the resistance and the conductance of the line are changing with frequency. Dispersion distorts the wave shape of the signal because the delay varies with frequency due to the inductance and capacitance of the line. The resistance, conductance, and capacitance are all distributed quantities along the length of the line. So the line can be represented by the equivalent circuit shown in Figure E.3.

One way to model a lossy transmission line is to use the circuit in Figure E.3. This type of modelling is called the lumped model approach, and it involves connecting a number of lumped line segments in series. The number of segments clearly depends on the line length and the simulation accuracy required. PSpice supports this type of lossy transmission lines with more than 40 commonly used coaxial cables models. These models are included in the "TLINE .LIB" PSpice library. It should be noted that PSpice allows the user to specify up to 128 lumped segments. A problem with the lumped approach is that to obtain accurate results, a large number of segments are often required, leading to long simulation time and large netlists. Analogue behavioural modelling (Chapter 8) allows an efficient method of modelling lossy transmission lines[1]. Here, PSpice uses a distributed model to represent the properties of a lossy transmission line, that is, the line resistance, inductance, conductance, and capacitance are all continuously apportioned along the line's length. This type of modelling is called the distributed model approach. The distributed model of a lossy transmission line is expressed in PSpice in the form of mathematical equations. This model is described in terms of five parameters as shown in Table E.1. The description statement of a lossy transmission line is

T<name> <A port + node> <A port – node> <B port + node>
  + <B port – node> LEN = <value> R = <value> L = <value> G = <value>
  + C = <value>

Table E.1 Model Parameters of Lossy Transmission

Model parameter	Description	Unit
R	Resistance per unit length	Ohm/m
L	Inductance per unit length	H/m
G	Conductance per unit length	mho/m
C	Capacitance per unit length	F/m
LEN	Electrical length	m

This statement is similar to the description of ideal transmission lines, but in the case of a lossy transmission line the electrical length, R, L, G, and C per unit length values, must be specified by the user. These values are obtained as follows: at high frequency, the characteristic impedance of a transmission line is expressed in terms of L and C

$$Z0 = \sqrt{\frac{L}{C}}$$

Usually the capacitance value per meter, C, is quoted by the manufacturer, as well as the value of Z0. From this information, the L value can be found. The parameters R and G of the lossy transmission line are functions of frequency, with R increases in proportion to the square root of frequency and G increases in proportion to the frequency. The coefficients of R and G are found experimentally or by solving simultaneous equations[2]. To illustrate the use of lossy transmission lines based on a distributed model, consider the following example.

Example E.1
Simulate the response of a 30m Belden 8281 coaxial cable. The PSpice input file of the circuit (Figure E.4) is given in listing E.1. It has been assumed that the cable is to operate between a source and load impedance of 75Ω.

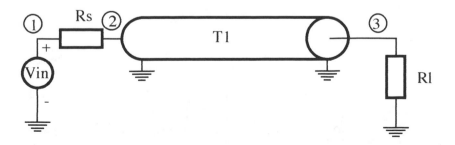

Figure E.4 Transmission line with source and load impedance=75Ω.

Listing E.1   PSpice Input File of Example E.1

```
Simulation of 30m of Belden 8281 Coaxial Cable
*
Vin 1 0 AC 1
.AC LIN 500 0.1E6 10E6
*
Rs 1 2 75
Rl 3 0 75
*
* modelling of 30m Belden 8281 cable
*
T1 2 0 3 0 LEN=30 R={78u*sqrt(s)} L=0.378u
+ G={1.6p*s} C=69p
*
.PROBE V(1), V(3)
*
.END
```

REF LEVEL   /DIV   MARKER 10 000 000.000Hz
-0.103dB   0.100dB   MAG (UDF)   -0.778dB

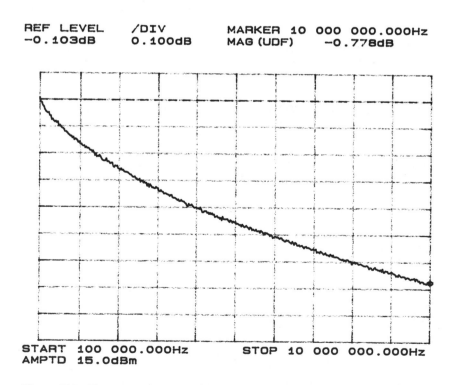

START 100 000.000Hz   STOP 10 000 000.000Hz
AMPTD 15.0dBm

Figure E.5   Frequency response (attenuation and group delay) of Example E.1.

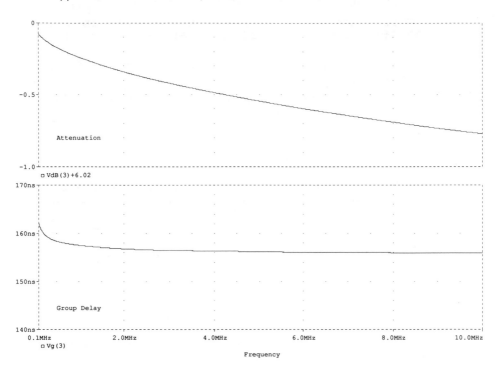

Figure E.6a Measured amplitude frequency response of a 30m Belden 8281 coaxial cable.

The manufacturer's data sheet indicates that Belden 8281 coaxial cable has $Z0 = 75\Omega$, and $C = 69\text{pF/m}$, which gives $L = 0.378\mu\text{H/m}$. The coefficients of R and G were found by iteration and measurement of two different lengths of cable. The model values of the Belden 8281 are shown in the listing. Note how R and G values are described using Laplace variable, s, expressions. Figure E.5 shows the frequency response of the simulated results, which compare well with the practical results shown in Figure E.6.

E.3 Conclusion

This appendix has shown that PSpice is capable of simulating ideal and lossy transmission lines. There are two models for lossy transmission lines, lumped and distributed, both of which are available in the "TLINE .LIB" PSpice library. The distributed model, however, provides faster simulation and more accurate results.

REF LEVEL   /DIV   MARKER 9 901 000.000Hz
150.00nSEC   5.000nSEC   DELAY (UDF)   152.21nSEC

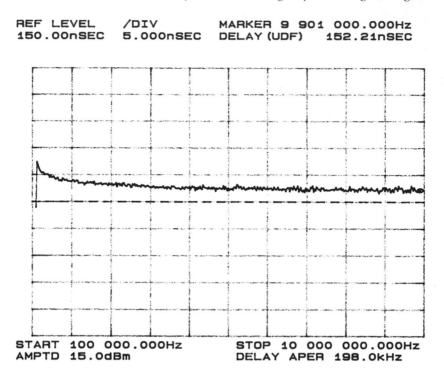

START 100 000.000Hz   STOP 10 000 000.000Hz
AMPTD 15.0dBm   DELAY APER 198.0kHz

Figure E.6b   Measured group delay response of a 30m Belden 8281 coaxial cable.

References

1. MicroSim Corporation, *PSpice Newsletter*, April 1991, pp. 9-12.
2. Chapman, R., *Transmission Lines*, McGraw-Hill, New York, 1968, pp. 65.

appendix F

PSpice control options

There are a number of options available that allow the user to control various aspects of a PSpice run. PSpice options are set using the .OP-TIONS statement. This statement, which must be specified in the circuit-input file, has the following basic form:

.OPTIONS [option name] [<option name> = <value>]

There are two types of options: those that do not require numerical values to be set and those that do.

F.1 PSpice options without values

These options generally allow the user to print and suppress circuit and simulation information as required. Table F.1 lists the options that can be selected or deselected. Options that begin with "NO" (e.g., NOECHO) will normally be printed by PSpice unless the user deselects them. The other options will not normally be printed, and the user must select them if required. For example

.OPTIONS NOECHO LIST

This statement instructs PSpice to suppress the listing of the circuit input file and to print out a summary of the circuit elements.

F.2 PSpice options requiring numerical values

These options allow the user to control various aspects of the simulation. PSpice has more than 30 options available[1]. The most important options are often considered to be those associated with accuracy of simulation. These options are shown in Table F.2. It is important to use these options carefully since they usually affect simulation time, accuracy, and memory requirements. Although all the parameters in Table F.2 control simulation accuracy, the parameter RELTOL is the most important since it determines the relative accuracy of all the calculated voltages and currents. The default value of RELTOL is 0.001, which is more accurate than is required for many applications. Sometimes it is

Table F.1   PSPice Options without Values

Option	Use the option to
NOBIAS	Suppress the printing of the bias point node voltages
NOECHO	Suppress listing of the input circuit file
NOMOD	Suppress listing of the model parameters and temperature updated values
NOPAGE	Suppress paging and the banner of each major section of output
NOREUSE	Suppress the automatic saving and restoring of bias point information between different temperature, Monte Carlo runs, worst-case runs, and parametric analyses (.STEP)
ACCT	Summarise accounting information at the end of all the analyses
EXPAND	List devices created by subcircuit expansion and list contents of the bias point file
NODE	List summary of node connection in the form of a table
LIBRARY	List lines used from library files
LIST	List summary of circuit elements
OPTS	List values for all options

Table F.2   PSpice Options with Values

Option	Meaning	Default	Unit
ABSTOL	Best accuracy of currents	1pA	A
CHGTOL	Best accuracy of charges	0.01pC	C
RELTOL	Relative accuracy of voltages and currents	0.001	—
VNTOL	Best accuracy of voltages	1 µV	V

necessary to decrease the accuracy of simulation by changing the value of RELTOL to 0.01 (see Appendix D, Listing D.1).

When using PSpice for high voltage and currents, it may be appropriate to increase the values of VNTOL and ABSTOL. Their default values are intended for circuits with voltages in the range of 1 to 10V and currents in the range of milliamps.

If the requirement is to simulate circuits that have voltages and currents in the range of kilovolts and amps, the values of the parameters VNTOL and ABSTOL need to be increased, for example, to 1mV and 1µA, respectively.

Reference

1. MicroSim Corporation, *Circuit Analysis — Reference Manual* (The Design Center), Version 5.3, Irvine, CA, January 1991.

appendix G

CMOS circuits simulation

PSpice is capable of simulating MOSFET transistors, including N and P channels. Two statements are required to describe MOSFET transistors, a component and .MODEL statement. The basic form of the MOSFET component statement is

M*<name>* *<ND>* *<NG>* *<NS>* *<NB>* *<model name>*
+ [L=*<value>*] [W=*<value>*] [AD=*<value>*]
+ [AS=*<value>*] [PD=*<value>*] [PS=*<value>*]

where M is the PSpice symbol for a MOSFET transistor and *<name>* is the name of the transistor. Examples are M1 and Mbias. The parameters *<ND>*, *<NG>*, *<NS>*, and *<NB>* are the drain, gate, source, and bulk or substrate nodes of the MOSFET transistor[1] as shown in Figure G.1. The parameter *<model name>* is usually a descriptive name of the transistor. This name can begin with any character and can be up to eight characters long. Examples of model names are TYPE1, N1, and P2. The optional parameters allow the user to describe an integrated circuit MOSFETs in terms of (L) and (W), which define the length and the width of the channel, respectively, in meters. The parameters (AD) and (AS) define the drain and the source area, respectively, in square meters. The parameters (PD) and (PS) define the perimeter of the drain and the source, respectively, in meters. If the user does not specify the optional parameters, PSpice will set the parameters (L and W) to 100μ, while setting the parameters (AD, AS, PD, and PS) to zero, the default values. In the early stages of simulation, only W and L are specified, since the other parameters cannot be specified until the transistor is geometrically defined.

The basic form of the .MODEL statement is

.MODEL *<model name>* *<type>* [model parameters]

The parameter *<model name>* is the name given to the MOSFET transistor specified in the component statement. The parameter *<type>* deter-

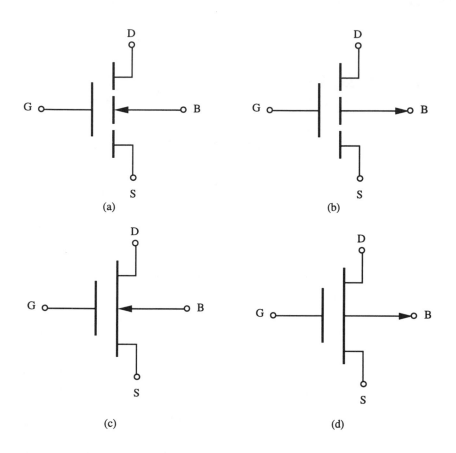

Figure G.1 MOS transistor symbols: (a) enhancement NMOS; (b) enhancement PMOS; (c) depletion NMOS; (d) depletion PMOS.

mines the MOSFET type (N or P channel), specified in PSpice as NMOS or PMOS, respectively. PSpice has three different MOSFET models, which are characterised by more than 40 parameters, some of which are shown in Table G.1.

The model parameter LEVEL specifies the required MOSFET model:

LEVEL=1 is the Shichman-Hodges model[2], which is the simplest, and also the default MOSFET transistor model, which is useful for simple circuit simulation.

LEVEL=2 is an advanced version of LEVEL 1 model, which takes into account the geometry of the MOSFET transistor as well as second-order effects[3].

Table G.1   MOSFET Transistor Model Parameters

Parameters	Meaning	Default value	Unit
LEVEL	Model type (1, 2, or 3)	1	
L	Channel length	DEFL	m
W	Channel width	DEFW	m
LD	Lateral diffusion (length)	0	m
WD	Lateral diffusion (width)	0	m
VTO	Zero-bias threshold voltage	0	V
KP	Transconductance	2E-5	A/V^2
GAMMA	Bulk threshold parameter	0	$V^{1/2}$
PHI	Surface potential	0.6	V^{-1}
LAMBDA	Channel-length modulation (LEVEL=1 or 2)	0	V
RD	Drain ohmic resistance	0	Ohm
RS	Source ohmic resistance	0	Ohm
RG	Gate ohmic resistance	0	Ohm
RB	Bulk ohmic resistance	0	Ohm
RDS	Drain-source shunt resistance	Infinite	Ohm
RSH	Drain, source diffusion sheet resistance	0	Ohm/ square
IS	Bulk p-n saturation current	1E-14	A
JS	Bulk p-n saturation current/area	0	A/m^2
PB	Bulk p-n potential	.8	V
CBD	Bulk-drain zero-bias p-n capacitance	0	F
CBS	Bulk-source zero-bias p-n capacitance	0	F
CJ	Bulk p-n zero-bias bottom capacitance/area	0	F/m^2
CJSW	Bulk p-n zero-bias perimeter capacitance/length	0	F/m
CGSO	Gate-source overlap capacitance/channel width	0	F/m
CGDO	Gate-drain overlap capacitance/channel length	0	F/m
CGBO	Gate-bulk overlap capacitance/channel length	0	F/m
KF	Flicker noise coefficient	0	
AF	Flicker noise exponent	1	

LEVEL=3 is an empirical model[3], which is designed for MOSFETs with short channels. Both LEVEL 2 and 3 models require a large amount of simulation time and could cause convergence problems. These models are usually used prior to circuit fabrication. A detailed discussion on the use of the various model levels regarding IC implementation is given in Reference 1.

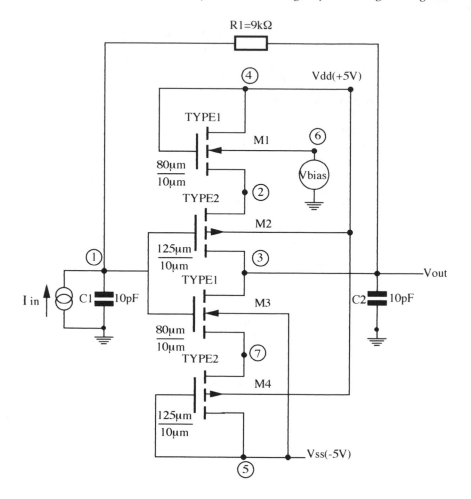

Figure G.2   Second-order lowpass OTA-based filter.

There are two modes of MOSFET transistors, enhancement and depletion. The model parameter VTO (threshold voltage) of the MOSFET transistor sets the mode in conjunction with the <*type*> part of the .MODEL statement. The parameter VTO is set positive for enhancement NMOS and for depletion PMOS. The parameter VTO is set negative for enhancement PMOS and for depletion NMOS. To illustrate the simulation of MOSFET transistors, consider the following example.

Listing G.1 PSpice Input File of Example G.1

```
Transconductance amplifier based filter (Figure G.2)
*
Iin 1 0 AC 0.21mA                   ; Current source
.AC DEC 10 10 10E6
Vdd 4 0 DC 5V
Vss 5 0 DC -5V
Vbias 6 0 DC -2V                    ; DC bias voltage
*
C1 1 0 10pF
C2 3 0 10pF
R1 1 3 9k
*
M1 4 4 2 6 TYPE1 W=80u L=10u     ; Enhancement NMOS transistor
M2 2 1 3 4 TYPE2 W=125u L=10u    ; Enhancement PMOS transistor
M3 3 1 7 5 TYPE1 W=80u L=10u     ; Enhancement NMOS transistor
M4 7 5 5 4 TYPE2 W=125u L=10u    ; Enhancement PMOS transistor
*
* CMOS transistors model parameters
*
.MODEL TYPE1 NMOS
+ vto=0.8 kp=44.2u gamma=0.951 lambda=0.04 phi=0.71
*
.MODEL TYPE2 PMOS
+ vto=-0.8 kp=15.2u gamma=0.344 lambda=0.04 phi=0.80
*
.PROBE V(3)
*
.END
```

Example G.1

Figure G.2 shows a second-order lowpass filter[4] (–3dB cutoff frequency=2.5MHz) based on a transconductance amplifier as the active element. The PSpice input file of the circuit is given in Listing G.1. The model name of the enhancement NMOS transistors (M1 and M3) is TYPE1, while the model name of the enhancement PMOS transistors (M2 and M4) is TYPE2. These names have been chosen arbitrarily. It has been assumed that the MOSFET transistors have LEVEL 1 models. The simulated filter frequency response is shown in Figure G.3.

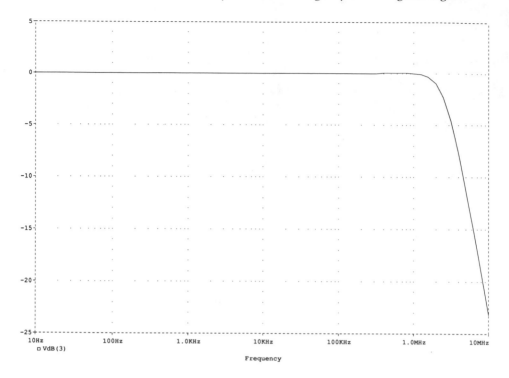

Figure G.3 Simulated frequency response of the circuit in Figure G.2.

Reference

1. Allen, P.E., & Holberg, D.R., *CMOS Analogue Circuit Design*, Holt, Rinehart and Winston, New York, 1987, pp. 134.
2. Shichman, H., & Hodges, D.A., "Modelling and Simulation of Insulated-Gate Field-Effect Transistor Switching Circuits", *IEEE Journal of Solid-State Circuits*, SC-3, September 1968, pp. 285.
3. Valdimirescu, A., & Liu, S. "The Simulation of MOS Integrated Circuits Using SPICE2", Memorandum No. UCB/ERL M8/7, College of Engineering, University of California, Berkeley, October 1980.
4. Al-Hashimi, B.M., & Fidler, J.K., "Novel High Frequency Continuous-Time Lowpass OTA Based Filters", IEEE International Symposium on Circuits and Systems, New Orleans, May 1990.

Index